MW00845889

Quantum Field Theory II

Quantum Field Theory II

Mikhail Shifman
University of Minnesota, USA

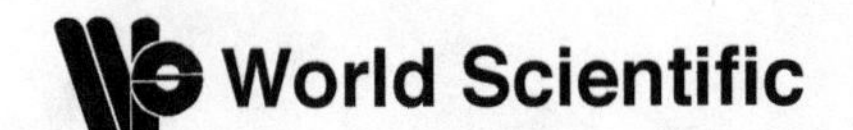

NEW JERSEY • LONDON • SINGAPORE • BEIJING • SHANGHAI • HONG KONG • TAIPEI • CHENNAI • TOKYO

Published by

World Scientific Publishing Co. Pte. Ltd.
5 Toh Tuck Link, Singapore 596224
USA office: 27 Warren Street, Suite 401-402, Hackensack, NJ 07601
UK office: 57 Shelton Street, Covent Garden, London WC2H 9HE

Library of Congress Cataloging-in-Publication Data
Names: Shifman, Mikhail A., author.
Title: Quantum field theory II / Misha Shifman (University of Minnesota, USA).
Description: Singapore ; Hackensack, NJ : World Scientific Publishing Co. Pte. Ltd., [2019] |
Includes bibliographical references and index.
Identifiers: LCCN 2019001401| ISBN 9789813234185 (hardcover ; alk. paper) |
ISBN 9813234180 (hardcover ; alk. paper)
Subjects: LCSH: Quantum field theory.
Classification: LCC QC174.46 .S553 2019 | DDC 530.14/3--dc23
LC record available at https://lccn.loc.gov/2019001401

British Library Cataloguing-in-Publication Data
A catalogue record for this book is available from the British Library.

For any available supplementary material, please visit
https://www.worldscientific.com/worldscibooks/10.1142/10825#t=suppl

Contents

Preface

The course of quantum field theory at the University of Minnesota is divided in three parts and lasts for three (or in lucky years, four) semesters. Somewhat symbolically, these parts are called Field Theory I, Field Theory II, and Field Theory III.

Field Theory III is not offered every year; it covers such topics as quantum field theories at strong coupling and basics of supersymmetry. Not too long ago I collected all my notes on Field Theory III and published them as a textbook [1].

Field Theory I treats relativistic quantum mechanics, spinors and the Dirac equation, introduces the Hamiltonian formulation of quantum field theory, and canonic quantization procedure. Then basic field theories (scalar, Yukawa, QED) are discussed, and Feynman's perturbation theory is worked out at the tree level. It usually ends with a brief survey of basic QED processes. Frequently used textbooks covering the above topics are F. Schwabl, *Advanced Quantum Mechanics*, (Springer, 1997) and F. Mandl and G. Shaw, *Quantum Field Theory*, (Second Edition, John Wiley and Sons, 2005). I also highly recommend T. Banks, *Modern Quantum Field Theory*, (Cambridge University Press, 2008) and D. Tong, *Quantum Field Theory* and *Lectures on Gauge Theory*.[1] The formalism I prefer to teach in the course of Field Theory I follows that introduced in [2], with a necessary modernization and/or some simplifications and amendments. My Field Theory I syllabus is attached below as Appendix on page xiv.

Field Theory II – the subject of the present course – begins with the path integral formulation of quantum field theory.[2] Perturbation

[1]http://www.damtp.cam.ac.uk/user/tong/qft/qft.pdf
http://www.damtp.cam.ac.uk/user/tong/gaugetheory.html

[2]Sometimes I find it useful to refresh students' memory of non-Abelian gauge theories before path integral formulation.

theory is generalized beyond tree level, to include radiative corrections (loops). The renormalization procedure and Wilsonean renormalization group are discussed, asymptotic freedom of non-Abelian gauge theories derived, and some applications in quantum chromodynamics (QCD) considered. I make a brief digression into the standard model (SM). Sample higher order corrections are worked out. The SM case requires a study of the spontaneous breaking of gauge symmetry, a phenomenon which would be more appropriate to call "Higgsing of the gauge bosons."

There are a number of modern textbooks which can be used for teaching Field Theory II. A typical text is that by M. Peskin and D. Schroeder [3]. Some chapters from [4; 5; 6; 6; 7; 8; 9], and [10] can be used as a supplement.[3] I do not think that the above textbooks can be used individually in their entirety: some are too long, others lack necessary technical details. I would not base my course of QFT II solely on [3] as it often happens, although I always advise my students to have this textbook handy. I also recommend [10] as a useful and entertaining warm-up reading before the beginning of the course.

That's why I collected my notes, organized them in more or less coherent form and assembled this relatively short textbook.[4] To my mind it contains everything which belongs to this course, and is not excessively complicated. Compared to Peskin and Schroeder it is much more concise. On the other hand, it also contains some examples of theories pertinent to condensed matter. Usually, in my class I have a mixed group of students: future theorists and experimentalists both in high energy physics and condensed matter. This is rather typical almost for all American universities. My task was not just to familiarize students with the basic concepts of QFT but, rather, to teach them how to carry out calculations in various problems that arise in quantum field theory. This is the main difference with the recently published text [10] and the well-known and popular volume [4] entitled *Quantum Field Theory in a Nutshell.*

The sections marked by asterisks are intended for especially curious readers. They are not always discussed in classes.

Lectures 22 and 23 (pages 229 and 239) rarely make it to this course due to the lack of time by the end of the semester. Usually I include

[3]The elements of the Lie group theory needed for this course will be discussed below in future lectures. For more details I refer the reader to [11].

[4]If you do not count Appendices on pages 239–298 (a supplementary lecture, a set of problems, and a brief history of quantum field theory) the lecture course *per se* is presented on $\sim$ 230 pages.

them in QFT III. I do not deliver in class *Brief History of QFT* presented in Appendix C, see page 275. At the same time I think that the knowledge of how quantum field theory had emerged is a must for every student taking the courses of quantum field theory. Therefore, I distribute *Brief History of QFT* as a handout before the beginning of the semester, and ask the students to read it at home before they come to class.

Each week my students are given a problem or two relevant to the topic of the week which they were supposed to solve at home. Most of these problems are included as exercises in Appendix C, page 249. The problems marked by asterisk were not mandatory. Students could use them for extra credit.

Minneapolis
May 2018

Appendix. Field Theory I Syllabus

L1. Quantization of harmonic oscillator in quantum mechanics using the operator formalism. Hamiltonian vs. Lagrangian. Action. Heisenberg, Schrödinger and interaction representations.

L2. Scalar complex free fields: second quantization. "Large Box" as an infrared regularization (spectrum discretization). Lagrangian $\to$ Hamiltonian.

L3. Scalar complex free fields: continuation. Noether theorem. Conserved current. "One particle per volume V" normalization. T product. Massive scalar propagator in momentum space. Where the poles in the p_0 complex plane lie (the Feynman $i\,\epsilon$ rule).

L4. Fourier transform and massless scalar propagator in the coordinate space. Second quantization of the real scalar field. Introduction of the interactions terms ϕ^3 and ϕ^4. What is weak coupling?

L5. Complex scalar field with the "Higgs potential". Spontaneous breaking of $U(1)$. Goldstone bosons.

L6. Spinors in four dimensions starting from brief review of spinors in three dimensions. Decomposition of $O(1,3)$. Dotted and undotted Weyl spinors. Transformation laws under $O(3)$ rotations and Lorentz boosts. Building all representations from the spinor representations.

L7. Weyl, Dirac and Majorana spinors. Dirac equation. Dirac and Majorana gamma matrices. Plane wave solutions.

L8. Gamma matrix traces and algebra. Exercises. γ^5.

L9. Dirac Lagrangian and Hamiltonian. Canonic quantization of ψ and $\bar{\psi}$. Verifying the "one particle per volume V" normalization.

L10. Scalar QED in the classical limit. Gauge symmetry. Conserved current and the masslessness of the photon. Physical degrees of freedom of the photon field A_μ.

A_0 can be expressed (non-locally) through currents. Coulomb interaction.

L11. Photon field strength tensor. Canonic quantization of A_μ and $F_{\mu\nu}$. Various gauges.

L12. Photon propagator in the Coulomb gauge. Transition to the Feynman and other gauges.

L13. The Higgs phenomenon. Reshuffling of degrees of freedom: $2+2 \to 3+1$. The mass of the Higgsed photon. Propagators.

L14. Domain walls in $\lambda\phi^4$ (real scalar field with Z_2 symmetry). Domain wall tension.

L15. Spinor QED. Quantization. Development of perturbation theory. P, T, and C symmetries.

L16. Tree Feynman graphs in ϕ^3 and ϕ^4 scalar QFT. Tree Feynman graphs in scalar QED.

L17. Scattering matrix. Generalities: cross-section, decay rates. Transition from amplitudes to Feynman amplitudes.

L18. Fermion density matrices. Phase spaces in simple cases.

L19. Compton scattering in Dirac QED.

L20. Compton scattering in Dirac QED, continuation.

L21. Some basic QED processes such as $e^+e^- \to \mu^+\mu^-$.

L22. Crossing symmetry. Unitarity of the S-matrix and its consequences (i.e. the optical theorem).

L23. Form factors and invariant amplitudes, including the magnetic part of the $ee\gamma$ vertex.

L24. Calculation of processes with massless fermions using the Green functions in the coordinate space.

L25. First ideas about radiative corrections (Feynman diagrams with loops). Anomalous magnetic moment of the electron.

L26. Calculation of the anomalous magnetic moment of the electron. Analytic continuation to Euclidean vs. the Feynman $i\epsilon$ rule.

L27. Feynman parametrization of the loop integrals. Calculation of the Feynman integrals. Completion of the anomalous moment calculation.

L28. Effective action and first ideas of the running coupling constant. Heuristic calculation of $e^2(\mu)$. Landau zero charge (infrared freedom).

References

[1] M. Shifman, *Advanced Topics in Quantum Field Theory*, (Cambridge University Press, 2012), 622 pp.

[2] V. B. Berestetskii, L. P. Pitaevskii and E. M. Lifshitz, *Quantum Electrodynamics*, (Landau-Lifshitz Course of Theoretical Physics), Second Edition, (Butterworth-Heinemann, 1982).

[3] M. Peskin and D. Schroeder, *An Introduction to Quantum Field Theory*, (Addison-Wesley, 1995).

[4] A. Zee, *Quantum Field Theory in a Nutshell*, (Princeton University Press, Princeton, 2003).

[5] C. Itzykson and J.-B. Zuber, *Quantum Field Theory*, (McGraw-Hill, 1980).

[6] M. Srednicki, *Quantum Field Theory*, (Cambridge University Press, Cambridge, 2007).

[7] Michele Maggiore, *A Modern Introduction to Quantum Field Theory*, (Oxford University Press, 2005).

[8] Ashok Das, *Lectures on Quantum Field Theory*, (World Scientific, 2008).

[9] T. Banks, *Modern Quantum Field Theory: A Concise Introduction*, (Cambridge University Press, Cambridge, 2008).

[10] A. Smilga, *Digestible Quantum Field Theory*, (Springer, 2017).

[11] H. Georgi, *Lie Algebras in Particle Physics: From Isospin to Unified Theories*, (CRC Press, 1999);

Pierre Ramond, *Group Theory: A Physicist's Survey*, (Cambridge University Press, 2010);

A. Zee, *Group Theory in a Nutshell for Physicists*, (Princeton University Press, 2016).

Chapter 1

Lecture 1. Yang-Mills Theory

> *Prerequisites: (a) Canonic quantization in ϕ^4 theory and QED; (b) tree Feynman graphs; (c) basics of Lie groups (mostly $SU(N)$), generators, representations – fundamental and adjoint – group structure constants.*

In the previous semester we covered basics and general aspects of quantum field theory. In this semester we will consider some particular, the most important field theories such as non-Abelian gauge theories (also known as Yang-Mills theories). During the last six decades, Yang-Mills theory has become the cornerstone of theoretical physics. It is seemingly the only fully consistent relativistic quantum field theory in four space-time dimensions. As such, it is the underlying theoretical framework for the Standard Model of Particle Physics (a part of which is the Glashow-Weinberg-Salam, GWS) model, which was proven to be the correct theory at all currently measurable energies. For recommended primary textbooks, see [1].

A few words are in order here as a warm-up introduction. Theoretical physics is an enormous subject, arguably, the most important fundamental science of nature. It is convenient to classify it using the so-called "magic $cG\hbar$ cube" invented and discussed in the late 1920s and early 1930s [2; 3]. It is shown in Fig. 1.1. Here c is the speed of light in vacuum. (Also, it is the maximal velocity of any object in nature.) It measures the extent of relativity. Next, G is the Newton constant. It normalizes gravity. Finally, analogously to c, the quantity $\hbar$ is another fundamental constant, the so-called Planck constant. It tells us when classical physics is overtaken by quantum physics. When you come to take this lecture course, you are supposed to already know all but two branches of theoretical physics indicated in Fig. 1.1.[1] The subject of my course is the phenomena which occur in systems with

[1] The back right upper corner is, perhaps, problematic.

typical velocities close to c and typical actions of the order of $\hbar$. This is the front right lower corner of the cube. Near the front right upper corner gravity effects become nonperturbative. This corner is supposed to be described by a future theory. Perhaps, it will be string theory, or something else, we do not know. And I will not venture into this territory.

In this course, as in its first part, I will use the system of units in which $c = \hbar = 1$. If so, energy and momentum have dimension of mass while length and spatial coordinates have dimension 1/mass. The Newton constant then defines a "fundamental" mass, also known as the Planck mass,

$$m_{\mathrm{P}} = \sqrt{\hbar c/G} \approx 1.22 \times 10^{19}\,\mathrm{GeV}\,, \tag{1.1}$$

or, given that $c = \hbar = 1$,

$$G = \frac{1}{m_{\mathrm{P}}^2}\,. \tag{1.2}$$

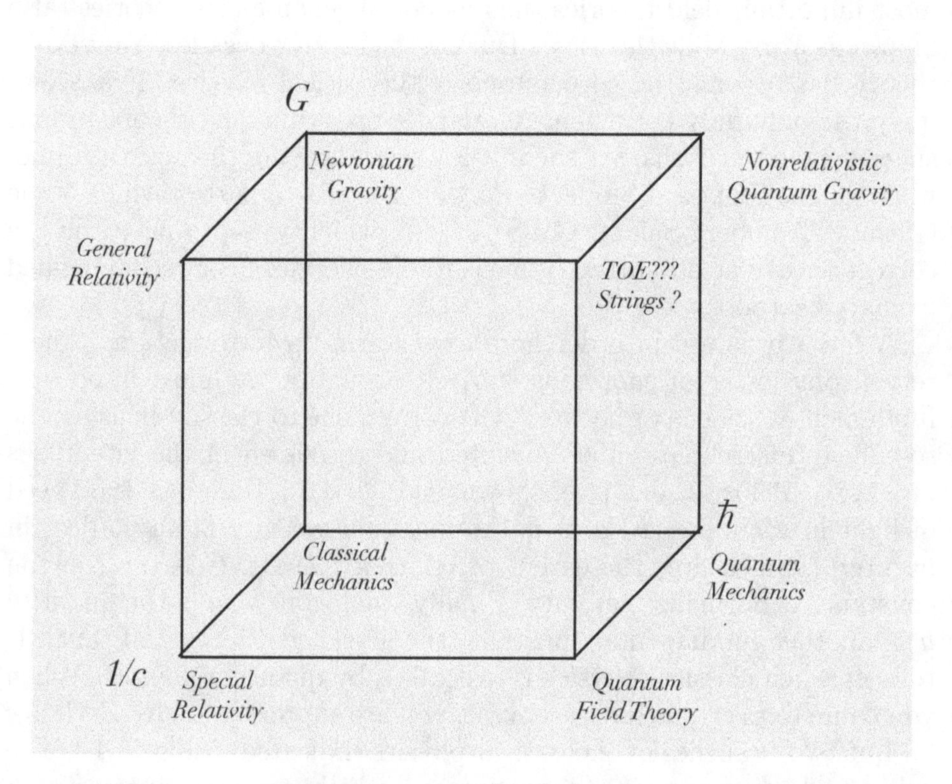

Fig. 1.1 The $cG\hbar$ cube of physics.

1.1 Construction of non-Abelian gauge theories

As you remember, gauge symmetry is *not* a symmetry but, rather, a redundancy in the description of the theory occurring when one elevates a global symmetry to the status of local symmetry. Let us review first the example of scalar QED (i.e. quantum electrodynamics of one scalar complex field ϕ).

Let us start with the globally U(1) invariant theory [2]

$$\mathcal{L} = (\partial_\mu \phi^\dagger)(\partial^\mu \phi) - V(|\phi|)\,, \tag{1.3}$$

where V is the potential,

$$V = -m^2 \phi^\dagger \phi + \frac{\lambda}{2}(\phi^\dagger \phi)^2\,. \tag{1.4}$$

This theory has a symmetry under U(1) rotations,

$$\phi \to e^{i\alpha}\,\phi\,, \qquad \alpha = \text{const.} \tag{1.5}$$

Correspondingly, the theory has a continuous vacuum manifold. Any point

$$\phi_{\rm vac} = e^{i\alpha}\,v \equiv e^{i\alpha}\,\frac{m}{\sqrt{\lambda}} \tag{1.6}$$

is a valid vacuum (ground state). Above I assumed that the parameters m and λ are real and positive.

Now we would like to make the above theory invariant under local transformations with the phase $\alpha(\vec{x}, t)$. The potential V is obviously invariant. However, the kinetic term is not. To make it invariant we must add the photon field A_μ, and replace the partial derivative by a covariant derivative, $\partial^\mu \to D^\mu$, such that ϕ and $D^\mu\phi$ transform in one and the same way, namely,

$$\phi_{\rm gt}(x) = e^{i\alpha(x)}\,\phi(x)\,, \qquad \left(D^\mu\phi(x)\right)_{\rm gt} = e^{i\alpha(x)}\,\left(D^\mu\phi(x)\right)\,, \tag{1.7}$$

where the subscript gt stands for gauged transformed, and the four-coordinate x^μ for brevity is written as x with the superscript omitted, $x \leftrightarrow \{t,\, \vec{x}\}$. The second equality in (1.7) can be viewed as a basic definition of the covariant derivative. It is obviously satisfied if

$$D^\mu = \partial^\mu - iA^\mu \tag{1.8}$$

and the field $A^\mu(x)$ transforms as

$$A^\mu_{\rm gt} = A^\mu + \partial^\mu \alpha(x)\,. \tag{1.9}$$

[2]The metric used throughout this text is $g^{\mu\nu} = \text{diag}\,(1, -1, -1, -1)$.

Combining (1.5), (1.8) and (1.9) it is very easy to check that (1.7) is valid. In the case at hand we certainly know (from the classical theory of electromagnetism) that the field A^μ which appeared in the process of "gauging"[3] is the electromagnetic four-potential. Its kinetic term is proportional to $F_{\mu\nu}F^{\mu\nu}$ where

$$F_{\mu\nu} = \partial_\mu A_\nu - \partial_\nu A_\mu \,. \tag{1.10}$$

Let us note the following identity relating $F_{\mu\nu}$ to a commutator[4] of covariant derivatives:

$$F^{\mu\nu} \equiv i\,[D^\mu D^\nu] \,. \tag{1.11}$$

It will help us carry out generalization to non-Abelian gauging.

Finally, the full Lagrangian takes the form

$$\mathcal{L} = -\frac{1}{4e^2}\, F_{\mu\nu}F^{\mu\nu} + (D_\mu\phi^\dagger)(D^\mu\phi) - V(|\phi|)\,, \tag{1.12}$$

where e is the coupling constant. From (1.8)–(1.10) it is clear that

$$\left(F^{\mu\nu}\right)_{\rm gt} = F^{\mu\nu}\,, \tag{1.13}$$

implying that the Lagrangian (1.12) is fully invariant under the local U(1) transformations presented in (1.7) and (1.8).

The phase of ϕ no longer presents a physical degree of freedom; rather, it is absorbed in photon's longitudinal polarization. (Remember, the photon acquires a mass provided the right-hand side in (1.6) does not vanish; this phenomenon is called Higgsing.) Simultaneously, the continuous vacuum manifold (1.6) shrinks into a single point since one can always choose, say, $\alpha = 0$ in (1.6) through a gauge condition on the fields. All points in (1.6) are gauge equivalent. There are no massless particles in the spectrum of the gauged theory with potential (1.4), in contradistinction with the ungauged theory (1.3).

In general, all filed configurations related to each other by gauge transformations represent one and the same point in the space of fields. This is why gauging a global symmetry introduces redundancy.

***

Following the above pattern let us generalize the idea of gauging to non-Abelian symmetries. Assume that the field ϕ in (1.7) now carries an index i,

$$\phi \to \phi^i\,, \tag{1.14}$$

[3]This is how the transition from global to local symmetry is referred to.

[4]It is assumed that the derivatives act on arbitrary complex functions.

where for definiteness[5] we will choose i to be the index of the fundamental representation of SU(N),

$$i = 1, 2, ..., N\,. \tag{1.15}$$

Then

$$\phi^\dagger \to \phi_i^\dagger\,, \tag{1.16}$$

belongs to the antifundamental representation.

The Lagrangian (1.3) takes the form

$$\mathcal{L} = (\partial\phi_i^\dagger)(\partial\phi^i) - V(\phi_i^\dagger\phi^i)\,. \tag{1.17}$$

It is easy to see that it is invariant under a *global* SU(N) transformation. Indeed, if

$$\phi \to U\phi\,, \quad \phi^\dagger \to \phi^\dagger U^\dagger\,, \tag{1.18}$$

where U is any constant unitary matrix with unit determinant, $U \in$ SU(N), then

$$\begin{aligned} &\phi_i^\dagger\phi^i \to \phi^\dagger U^\dagger\, U\,\phi \to \phi^\dagger\phi\,; \\ &(\partial\phi_i^\dagger)(\partial\phi^i) \to (\partial\phi^\dagger)U^\dagger U(\partial\phi) \to (\partial\phi^\dagger)(\partial\phi)\,. \end{aligned} \tag{1.19}$$

The invariance of the Lagrangian (1.17) is obvious.

Now we want to make SU(N) local. Note that matrix $U \in$ SU(N) can be represented as

$$U = \exp\left(i\omega^a T^a\right), \tag{1.20}$$

where T^a are the generators of SU(N) and ω^a are arbitrary parameters which may or may not be x dependent, $a = 1, 2, ..., N^2 - 1$. For SU(2) the generators in the fundamental representation are proportional to the Pauli matrices, and for SU(3) proportional to the Gell-Mann matrices. In both cases the proportionality coefficient is $1/2$. The standard normalization of the generators in the fundamental representation is

$$\mathrm{Tr}\left(T^a T^b\right) = \frac{1}{2}\delta^{ab} \tag{1.21}$$

for any SU(N). The defining commutation relations for the generators (in any representation) are

$$[T^a T^b] = i f^{abc} T^c\,, \tag{1.22}$$

[5]The procedure is absolutely general and works for any representation of any non-Abelian group.

where f^{abc} are the group structure constants. For SU(2) one has $f^{abc} \equiv \varepsilon^{abc}$, where ε^{abc} is the Levi-Civita antisymmetric tensor.[6]

The global invariance implies $U =$ const by definition, which means, in turn that all ω^as in (1.20) are x independent. Now we want U to depend on the space-time point, $U \to U(x)$. Correspondingly,

$$\omega^a \to \omega^a(x)\,, \qquad a = 1, 2, ..., N^2 - 1\,. \tag{1.23}$$

The potential term in (1.17) remains invariant under the local SU(N) transformation. We want to generalize the kinetic term to be invariant under any local (x-dependent) transformation.

To this end we need to define the covariant derivative in such a way that, as in (1.7), after the gauge transformation (1.18),

$$\left(D^\mu \phi(x)\right)_{\rm gt} = U(x)\,\left(D^\mu \phi(x)\right)\,. \tag{1.24}$$

The solution to this equation is more contrived than in the Abelian theory because of non-commutativity of different matrices $U(x) \in$ SU(N).

Let us assume (the assumption to be justified *a posteriori*) that

$$D_\mu = \partial_\mu - i A_\mu^a T^a\,, \tag{1.25}$$

where A_μ^a are non-Abelian gauge fields called gauge bosons (analogs of the electromagnetic four-potential), and T^as represent N^2-1 generators of SU(N). In QCD the A_μ^a fields are called gluons, for historical reasons.

Equation (1.24) is satisfied provided that

$$A^{\mu\,a}_{\rm gt}(x)\,T^a = \underbrace{U(x)\left(A^{\mu\,a}(x)\right)T^a\,U^\dagger(x)}_{\text{homogenious term}} - \underbrace{i\left(\partial^\mu U(x)\right)U^\dagger(x)}_{\text{inhomogenious}}\,. \tag{1.26}$$

Indeed,

$$\begin{aligned}
\left(D^\mu \phi(x)\right)_{\rm gt} &= \partial^\mu\left(U(x)\phi(x)\right)\\
&-i\left[U(x)\left(A^{\mu\,a}(x)\right)T^a\,U^\dagger(x) - i\left(\partial^\mu U(x)\right)U^\dagger(x)\right]\left(U(x)\phi(x)\right)\\
&= \left(\partial^\mu U(x)\right)\phi(x) + U(x)\partial^\mu\phi(x) - iU(x)\left(A^{\mu\,a}(x)\right)T^a\phi(x)\\
&-\left(\partial^\mu U(x)\right)\phi(x)\\
&= U(x)\left[\partial^\mu\phi(x) - i\left(A^{\mu\,a}\right)T^a\phi(x)\right] = U(x)D^\mu\phi(x)\,.
\end{aligned} \tag{1.27}$$

---

[6]The Levi-Civita tensor is also called permutation symbol or totally antisymmetric symbol. Tullio Levi-Civita (1873-1941) was a famous Italian mathematician.

For what follows it is useful to note that if $\omega^a \ll 1$, i.e. the gauge matrix $U(x)$ is close to unity (see (1.20)) then the change of $A^{\mu\,a}$ under the gauge transformation is

$$\delta A^{\mu\,a} = \partial^\mu \omega^a + f^{abc} A^{\mu\,b} \omega^c = D^\mu \omega^a\,. \tag{1.28}$$

In non-Abelian gauge theories the product $(A^{\mu\,a})\,T^a$ is often written as A^μ. In this shorthand the SU(N) index a is hidden. One must presume its presence from the context.

Examining Eq. (1.26) we note that, in contradistinction with the Abelian gauge transformation, $A^{\mu\,a}_{\rm gt}(x)\,T^a$ contains two terms: a non-homogeneous term (the second term on the right-hand side) analogous to what we had in the Abelian case, and an extra homogeneous term specific for non-Abelian gauge theories.

Finally, we have to establish the kinetic term for the gauge bosons. We will proceed analogously to Eq. (1.11). Assuming the covariant derivatives that act on an arbitrary column of N complex functions, we obtain

$$\begin{aligned} i\,[D^\mu D^\nu] &= \partial^\mu A^\nu - \partial^\nu A^\mu - i\,[A^\mu A^\nu] \stackrel{\rm def}{=} G^{\mu\nu} \\ &= \left[\partial^\mu A^{\nu\,a} - \partial^\nu A^{\mu\,a} + f^{bca} A^{\mu\,b} A^{\nu\,c}\right] T^a \end{aligned} \tag{1.29}$$

The combination in the square brackets is referred to as the gauge field strength tensor, in analogy with the electromagnetic field strength tensor,

$$G^{\mu\nu\,a} \equiv \partial^\mu A^{\nu\,a} - \partial^\nu A^{\mu\,a} + f^{bca} A^{\mu\,b} A^{\nu\,c}\,. \tag{1.30}$$

The nonlinear term on the right-hand side is absent in the Abelian case; it is due to noncommutativity of the group generator matrices.

What is the gauge transformation of the gluon field strength tensor? It is easy to derive it from Eq. (1.29) taking into account that

$$D^\mu_{\rm gt} = U D^\mu U^\dagger\,. \tag{1.31}$$

Then we find that

$$G^{\mu\nu}_{\rm gt} = U G^{\mu\nu} U^\dagger\,, \tag{1.32}$$

which implies in turn that $\mathrm{Tr}(G^{\mu\nu}\,G_{\mu\nu})$ is gauge invariant.

As a result, the kinetic term of the non-Abelian gauge field theory can be written as

$$\mathcal{L} = -\frac{1}{2g^2}\,\mathrm{Tr}\,(G^{\mu\nu}\,G_{\mu\nu}) = -\frac{1}{4g^2}\,(G^{\mu\nu\,a}\,G^a_{\mu\nu})\,, \tag{1.33}$$

where g is the gauge coupling constant. The full Lagrangian including the gauge and matter fields is

$$\mathcal{L} = -\frac{1}{2g^2}\,\mathrm{Tr}\,(G^{\mu\nu}\,G_{\mu\nu}) + (D^\mu\phi_i)^\dagger(D_\mu\phi^i) - V(\phi_i^\dagger\phi^i)\,. \tag{1.34}$$

Equation (1.34) includes all relevant operators. I hasten to add, however, that beyond perturbation theory we should add one extra term (which is P and T odd), namely the so-called θ term,

$$\mathcal{L}_\theta = \frac{\theta}{16\pi^2}\mathrm{Tr}\left(G^{\mu\nu}\,\tilde{G}_{\mu\nu}\right) = \frac{\theta}{32\pi^2}\left(G^{\mu\nu\,a}\,\tilde{G}^a_{\mu\nu}\right), \tag{1.35}$$

where

$$\tilde{G}^a_{\mu\nu} = \frac{1}{2}\varepsilon_{\mu\nu\alpha\beta}\tilde{G}^{\alpha\beta\,a} \tag{1.36}$$

is the dual field strength tensor. This term introduces a new parameter θ also known as the vacuum angle. Why the θ term is important only beyond perturbation theory and where it comes from is a separate (*albeit* important) topic to which we will turn much later, see page 240.

1.2 Fermion (quark) matter

In the above construction for pedagogical purposes I used the scalar field in the fundamental representation. In theories relevant to nature, as a rule, the matter sector consists of fermions. The fermion part of the Lagrangian is basically the same as in QED, with the exception of definition of the covariant derivative (see (1.25)),

$$\mathcal{L}_{\mathrm{ferm}} = \sum_f \bar{\psi}_f\,(i\,\gamma^\mu D_\mu - m_f)\,\psi^f\,, \tag{1.37}$$

where f is the flavor index, ψ is the Dirac fermion and m_f is the mass of the quark of flavor f in the representation R of the gauge group G. Equation (1.37) is written for quantum chromodynamics; it is assumed that all fermion fields are in the fundamental representation of the gauge group SU(N) (in actuality, SU(3)), and so are the generators of the gauge group in the covariant derivatives acting on the fermions. Should we anticipate generalizations?

The answer is positive. First, in the Standard Model we deal with the Weyl rather than Dirac fermions. This will be discussed in due time.

Second, the matter fields need not be necessarily in the fundamental representation. Generally speaking, we can consider matter fields in any representation. The only change is that in Eq. (1.25) for the covariant derivative acting on the given matter field we must take the generator matrices T^a in the appropriate representation. For instance, in supersymmetric Yang-Mills theory the fermions to be considered are in the adjoint representation.

For real representations of the gauge group G (e.g. the adjoint representation) the fermion fields in (1.37) can be Majorana fields.

1.3 Yukawa couplings

By definition, the Yukawa coupling is a three-field coupling: two of the fields involved are fermionic and one is a boson spin-zero field. Yukawa couplings exist only in special cases when the matter sector of the Yang-Mills theory at hand contains such fields that one can built a Lorentz-scalar gauge invariant operator from three fields. The mass dimension of this term must be four (or D in the general D-dimensional space). For instance, let us consider the SU(N) gauge theory with the Dirac fermions in the fundamental representation and a complex scalar field Φ^a in the adjoint representation. Then one could add to (1.34), (1.37) the Yukawa term

$$\mathcal{L}_{\text{Yukawa}} = h \left(\bar{\psi}\, T^a \psi\right) \Phi^a \,, \tag{1.38}$$

where T^a's are the generators of the SU(N) group in the fundamental representation and h is a Yukawa coupling constant. Another popular option is $\bar{\psi}\psi\,\Phi$ where Φ is a gauge singlet field. I advise you to play with various representations of fermions and ϕ fields to build a variety of Yukawa terms. You may also refresh your memory of the Yukawa terms in the Glashow-Weinberg-Salam theory.

Appendix 1.1: C. N. Yang and Robert Mills

Yang and Mills had developed Yang-Mills theory (in 1954) in the context of an attempt to describe the strong interactions of vector mesons, such as ρ mesons. The SU(2) gauge theory they found did not work for this purpose, since (as we now know) what was needed was an SU(3) theory of colored quarks and gluons which came only 20 years later.

C. N. Yang (1922 -) and Robert Mills (1927 - 1999)
at Stony Brook in 1999.

Fig. 1.2 In 1957 the Nobel Prize in Physics was awarded jointly to Chen Ning Yang and Tsung-Dao (T.D.) Lee "for their penetrating investigation of the so-called parity laws which has led to important discoveries regarding the elementary particles." Robert Laurence Mills (1927-1999) was a physicist, specializing in quantum field theory, the theory of alloys, and many-body theory. While sharing an office at Brookhaven National Laboratory, in 1954, Yang and Mills proposed what is now called Yang-Mills field. Their original goal was to describe ρ mesons as gauge bosons (i.e. gauge the isotopic symmetry of strong interactions). Mills became Professor of Physics at the Ohio State University in 1956, and remained there until his retirement in 1995.

In 1953, Pauli was interested in a six-dimensional theory of Einstein's field equations of general relativity along the lines suggested by Kaluza and Klein in five dimensions. He compactified two extra dimensions into two-dimensional sphere, which inevitably led him to SU(2) Yang-Mills theory. However, non-Abelian gauge bosons remain massless, and at that time the only massless fields known to physicists were photons and gravitons, plus the neutrino postulated by Pauli in 1931, and not yet discovered in 1953. Somewhere I read that, when asked why he did not publish his research, Pauli said: "I have already introduced one hypothetical massless particle, and I had no nerve to introduce more..." (see [4], Chapter 1).

Because of his super-high requirements for his own work in physics, he put on hold publication on Yang-Mills, see below. In the meantime the theory was independently developed by C. N. Yang and Robert Mills, reported at a Princeton seminar and published in *Physical Review.*

Yang recollects of a seminar he gave in Princeton where Pauli was very critical and Pais was also present. Pauli was asking Yang about the mass of the intermediate vector mesons (now gluons), probably knowing that they were massless and therefore a killer for the theory (there are no massless hadrons). Yang responded he wasn't sure of the answer. Apparently, Pauli was so insistent and hostile with his questions that Yang just sat down at the front row and stopped talking! Then Oppenheimer encouraged him to continue delivering his talk, which he did.

Pauli described his SU(2) version of Yang-Mills theory, before Yang and Mills, in the letter to Abraham Pais [1] (page 171), entitled "Meson Nucleon Interaction and Differential Geometry" (written "to see what it looks like," in three days in July, (July 22-25 1953). See also N. Straumann, [6].

References

[1] L. D. Faddeev and A. A. Slavnov, *Gauge Fields: An Introduction To Quantum Theory,* Second Edition, (CRC Press, 1993);
Pierre Ramond, *Field Theory: A Modern Primer*, Second Edition, (Addison-Wesley, 1990).

[2] G. Gamov, D. Ivanenko and L. Landau, Zh. Russ. Fiz. Khim. Obstva. Chast Fiz. **60**, 13 (1928), (in Russian).

[3] M. Bronshtein, *K voprosu o vozmozhnoy teorii mira kak tselogo* [On a possible theory of the world as a whole], in *Osnovnye problemy kosmicheskoy fiziki* [*Basic problems of cosmic physics*], Kiev, ONTI (1934), pp. 186-218, and, in particular p. 210 (in Russian); M. Bronshtein, Physikalische Zeitschrift der Sowjetunion, **9**, 140 (1936).

[4] M. Shifman, *Standing Together in Troubled Times: Unpublished Letters by Pauli, Einstein, Franck and Others*, (World Scientific, 2017).

[5] O'Raifeartaigh, *Dawning of Gauge Theory*, (Princeton University Press, 1997).

[6] N. Straumann, *On Pauli's Invention of non-Abelian Kaluza-Klein Theory in 1953*, http://arxiv.org/pdf/gr-qc/0012054.pdf

Chapter 2

Lecture 2. Yang-Mills Theories (*Continued*)

In this lecture we discuss vertices in Yang-Mills theory, and non-Abelian Higgs mechanism. As an example, we consider an SU(2) sector of the GWS model.

2.1 Vertices in Yang-Mills with quarks

To obtain the interaction vertices we start from the Lagrangian, expand it in powers of fields, discard terms quadratic in the fields (which, in turn, determine propagators, or Green's functions) and then examine the cubic and quartic terms. The propagators will be studied later, since the gauge fixing procedure in non-Abelian gauge theories is more subtle than in QED (Abelian theory). Now we start from vertices. The full Lagrangian is

$$\mathcal{L} = -\frac{1}{4g^2}\left(G^{\mu\nu\,a}\,G^a_{\mu\nu}\right) + \sum_f \bar{\psi}_f\left(i\,\gamma^\mu D_\mu - m_f\right)\psi^f\,, \qquad (2.1)$$

where f is the flavor index, ψ is the Dirac fermion and m_f is the mass of the quark of flavor f. In perturbation theory the kinetic term must be canonically normalized. To this end, as we know from QED, the coupling constant $1/g$ must be absorbed in the field A^a_μ. Then

$$\mathcal{L} = -\frac{1}{4}\left(G^{\mu\nu\,a}\,G^a_{\mu\nu}\right) + \sum_f \bar{\psi}_f\left(i\,\gamma^\mu D_\mu - m_f\right)\psi^f\,, \qquad (2.2)$$

where

$$G^{\mu\nu\,a} \equiv \partial^\mu A^{\nu\,a} - \partial^\nu A^{\mu\,a} + g\,f^{bca}\,A^{\mu\,b}A^{\nu\,c}\,, \qquad (2.3)$$

and

$$iD_\mu = i\partial_\mu + gA^a_\mu T^a\,. \qquad (2.4)$$

Discarding the quadratic terms we arrive at

$$\mathcal{L}_3 + \mathcal{L}_4 = -g\,(\partial^\mu A^{\nu\,a})\, f^{abc} A^{\mu\,b} A^{\nu\,c}$$
$$-\frac{g^2}{4} f^{abc} A^{\mu\,b} A^{\nu\,c} f^{ade} A^d_\mu A^e_\nu + g\bar{\psi}\gamma^\mu A^a_\mu T^a \psi\,. \tag{2.5}$$

Taking into account that the vertices are determined by $i\mathcal{L}$ and combinatorics we arrive at the following set of vertices:

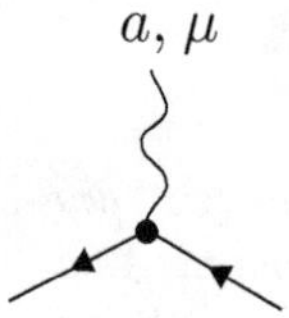

$$= ig\gamma^\mu T^a$$

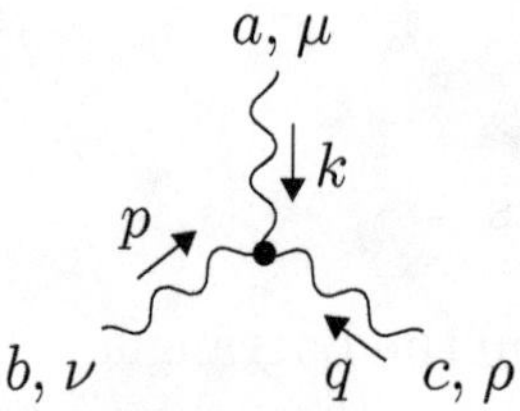

$$= g f^{abc}[g^{\mu\nu}(k-p)^\rho + g^{\nu\rho}(p-q)^\mu + g^{\rho\mu}(q-k)^\nu]$$

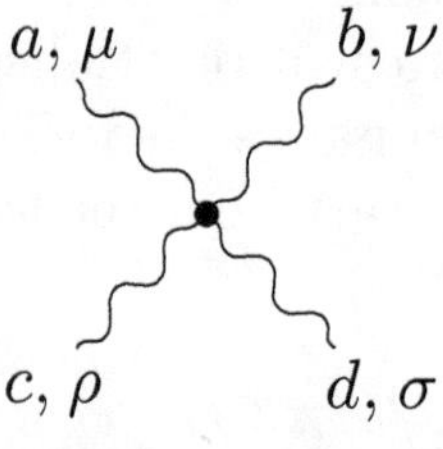

$$= -ig^2[f^{abc} f^{cde}(g^{\mu\rho}g^{\nu\sigma} - g^{\mu\sigma}g^{\nu\rho}) + f^{ace} f^{bde}(g^{\mu\nu}g^{\rho\sigma} - g^{\mu\sigma}g^{\nu\rho}) + f^{ade} f^{bce}(g^{\mu\nu}g^{\rho\sigma} - g^{\mu\rho}g^{\nu\sigma})]$$

Fig. 2.1 The set of vertices in Yang-Mills theory with fermions.

2.2 Higgsing

In the past, the Higgs mechanism was usually referred to as *spontaneous gauge symmetry breaking.* This was misleading since gauged symmetries are not at all physical symmetries and, moreover, they were not broken. Rather, they represent a special type of redundancy necessary in the consistent description of spin-1 fields. Currently, the phenomenon is known as *Higgsing* – a process in which gauge bosons acquire masses by "absorbing" certain scalar fields which become longitudinal components of the "Higgsed" gauge bosons. A rather general discussion of this process can be found in Chapter 11 of [1]. In this course we will limit ourselves to some particular examples which are most frequently used.

The SU(2) gauge theory with one complex scalar field plays a special role because it is a part of the Glashow-Weinberg-Salam (Standard) Model. Here we will use it to examine the pattern of Higgsing of this model, as an instructive example. We will see that all three gauge bosons in this model acquire equal masses.[1] This is not accidental. The equality of masses is explained by the fact that this model has, in addition to $\mathrm{SU}(2)_{\mathrm{gauge}}$, another symmetry, $\mathrm{SU}(2)_{\mathrm{global}}$ which is rarely discussed for reasons I do not understand.

The structure constants in the SU(2) group are

$$f^{abc} = \varepsilon^{abc}\,. \tag{2.6}$$

For the time being we will ignore fermions. Then the matter sector consists of an SU(2) doublet of complex scalar fields ϕ^i where $i = 1, 2$ (the Higgs field). In other words, ϕ^i are "scalar quarks" in the fundamental representation. The covariant derivative acts on ϕ^i as follows:

$$\mathcal{D}_\mu \phi(x) \equiv \left(\partial_\mu - iA_\mu^a T^a\right)\phi\,, \qquad T^a = \frac{1}{2}\tau^a\,, \tag{2.7}$$

where τ^a's are the Pauli matrices. We will choose the ϕ self-interaction potential in the form

$$V(\phi) = \frac{\lambda}{2}\left(\bar{\phi}\phi - v^2\right)^2\,, \tag{2.8}$$

where v is a real positive constant parameter.

Quite often people say that such a theory has just the SU(2) gauge symmetry. I will now demonstrate the existence of an additional global symmetry, so that the overall symmetry is

$$\mathrm{SU}(2)_{\mathrm{gauge}} \times \mathrm{SU}(2)_{\mathrm{global}}\,. \tag{2.9}$$

One can prove this in a number of ways. Probably, the fastest proof is as follows. Let us introduce a 2×2 matrix

$$X = \begin{bmatrix} \phi^1 & -\left(\phi^2\right)^* \\ \phi^2 & \left(\phi^1\right)^* \end{bmatrix}. \tag{2.10}$$

[1] In the Standard Model this is not the case because of the additional U(1) gauge boson.

The Lagrangian of the model rewritten in terms of X takes the form [2]

$$\mathcal{L} = -\frac{1}{4g^2}\, G^a_{\mu\nu}\, G^{\mu\nu,\,a} + \frac{1}{2}\mathrm{Tr}\,(D_\mu X)^\dagger\,(D^\mu X) - \frac{\lambda}{2}\left(\frac{1}{2}\,\mathrm{Tr}\,X^\dagger X - v^2\right)^2 . \tag{2.11}$$

Note that the generators T^a in the covariant derivative D act on the matrix X through the matrix multiplication from the left. This Lagrangian is obviously invariant under the transformation

$$X(x) \to U(x)X(x)M^\dagger\,, \tag{2.12}$$

supplemented by the appropriate transformation of the gauge fields related to the matrix $U(x)$. Here M is an arbitrary x-independent matrix from $\mathrm{SU}(2)_{\mathrm{global}}$. The symmetry (2.9) is apparent in Eq. (2.11). In the vacuum $\frac{1}{2}\mathrm{Tr}\,X^\dagger X = v^2$. Using gauge freedom (three gauge parameters) one can always choose the unitary gauge [2] in which the vacuum value of X is

$$X_{\mathrm{vac}} = v\begin{pmatrix}1 & 0\\ 0 & 1\end{pmatrix}. \tag{2.13}$$

This vacuum expectation value breaks both groups, $\mathrm{SU}(2)_{\mathrm{gauge}}$ and $\mathrm{SU}(2)_{\mathrm{global}}$, but the diagonal global $\mathrm{SU}(2)$ corresponding to $U = M$ remains unbroken. Thus, the symmetry breaking pattern is

$$\mathrm{SU}(2)_{\mathrm{gauge}} \times \mathrm{SU}(2)_{\mathrm{global}} \to \mathrm{SU}(2)_{\mathrm{glob.\,diag}}\,. \tag{2.14}$$

Three would-be Goldstone bosons [3] are eaten up by the gauge bosons transforming the latter into massive W bosons belonging to the triplet (adjoint) representation of the unbroken $\mathrm{SU}(2)_{\mathrm{diag}}$. There are no massless particles in this model. The physically observable excitations are three W bosons with mass $m_W = gv/\sqrt{2}$ and one Higgs particle (a single real field) corresponding to excitations of the singlet field proportional to $\mathrm{Tr}\,X^\dagger X$ with mass $\sqrt{2\lambda}\,v$.

Let us perform a simple counting. The doublet Higgs field ϕ^i has four real degrees of freedom. One of them, $\phi^\dagger_i\phi^i$, is analogous to $|\phi|$ in the U(1) theory. Therefore, it will represent the physical Higgs field. We are left with three real degrees of freedom (analogous to the phase α of the U(1) theory), which will be eaten up by three gauge bosons of $\mathrm{SU}(2)_{\mathrm{gauge}}$ becoming longitudinal components of three W bosons.

[2]See page 114.

[3]The Goldstone bosons are also called the Nambu-Goldstone bosons, after Yoichiro Nambu and Jeffrey Goldstone.

In the simplest cases the balance of the degrees of freedom allows one to determine the pattern of the symmetry breaking. Indeed, let us consider SU(3) Yang-Mills theory. The scalar sector will consist of one complex fundamental field ϕ^i where $i = 1, 2, 3$. We have now eight gauge bosons and six real components residing in ϕ^i. Again, $\phi_i^\dagger \phi^i$ is analogous to $|\phi|$ in the U(1) theory. Therefore, one real degree of freedom becomes the physical Higgs particle. We are left with five real degrees of freedom in ϕ^i. Let us assume that all of them are eaten up in Higgsing and become longitudinal components of five W bosons. We are left with three massless gauge bosons, the number of gauge bosons in SU(2). Thus, we conclude that in this case the pattern of the gauge symmetry breaking is

$$\mathrm{SU}(3) \to \mathrm{SU}(2)\,.$$

Let us verify this conclusion. By judiciously choosing the axes in the isospace one can always align the triplet field in the vacuum as

$$\phi_{\text{vac}} = \begin{pmatrix} v \\ 0 \\ 0 \end{pmatrix}, \tag{2.15}$$

where v is a real vacuum expectation value. In this frame, it is obvious that the first five Gell-Mann matrices act nontrivially, while λ_6, λ_7 and $\sqrt{3}\lambda_8 - \lambda_3$ act trivially.

2.3 From Minkowski to Euclidean formulation

In the next lecture we will start acquainting ourself with the path-integral formulation of quantum field theory. In many path-integral derivations it will be convenient to abandon the Minkowski formulation of the theory and perform the so-called Wick rotation to Euclidean space.[4]

In the Minkowski space one distinguishes between the contravariant and covariant vectors, v^μ and v_μ, respectively, $(\mu = 0, 1, 2, 3)$. The spatial vector $\vec{v}$ coincides with the components of the contravariant four-vector v^μ,

$$\vec{v} = \{v^1, v^2, v^3\}\,.$$

[4]Gian Carlo Wick (1909-1992) was an Italian theoretical physicist who made important contributions to quantum field theory, e.g. Wick's theorem.

In the Euclidean space the distinction between the lower and upper vectorial indices is immaterial, we consider just one vector $\hat{v}_\mu$ ($\mu = 1, 2, 3, 4$).[5]

As we will see later, the following rules of transition to Euclidean space apply. The Minkowski time t is replaced by a Euclidean time τ,

$$t \to -i\tau\,, \qquad \frac{d}{dt} \to i\frac{d}{d\tau}\,. \tag{2.16}$$

The spatial coordinates are not changed in passing to Euclidean, $\hat{x}_i = x^i$. According to (2.16),

$$\hat{x}_4 \equiv \tau = ix_0\,. \tag{2.17}$$

Clearly, when x_0 is continued to imaginary values, the zeroth component of the vector potential A_μ also becomes imaginary.

We define the Euclidean vector potential $\hat{A}_\mu$ as follows:

$$A^i = -\hat{A}_i\,, \qquad A_0 = i\hat{A}_4. \tag{2.18}$$

With this definition, the quantities $\hat{A}_\mu$, ($\mu = 1, \ldots, 4$) form a Euclidean vector and

$$\int A_\mu dx^\mu = \int \hat{A}_\mu d\hat{x}_\mu\,. \tag{2.19}$$

For the operator of covariant differentiation

$$D_\mu = \partial_\mu - iA^a_\mu T^a, \tag{2.20}$$

we obtain

$$D^i = -\hat{D}_i, \quad D_0 = i\hat{D}_4, \qquad \hat{D}_\mu = \frac{\partial}{\partial\hat{x}_\mu} - i\hat{A}^a_\mu T^a\,, \tag{2.21}$$

to be compared with (1.25). For the field strength tensor $G_{\mu\nu}$ we get

$$G^a_{ij} = \hat{G}^a_{ij}\,, \qquad G^a_{0j} = i\hat{G}^a_{4j}, \tag{2.22}$$

where the Euclidean field strength tensor $\hat{G}^a_{\mu\nu}$ is defined as follows:

$$\hat{G}^a_{\mu\nu} = \frac{\partial}{\partial\hat{x}_\mu}\hat{A}^a_\nu - \frac{\partial}{\partial\hat{x}_\nu}\hat{A}^a_\mu + f^{abc}\hat{A}^b_\mu\hat{A}^c_\nu\,. \tag{2.23}$$

It is expressed in terms of $\hat{A}_\mu$ and $\partial/\partial\hat{x}_\mu$ just in the same way as the Minkowskian $G^a_{\mu\nu}$ in terms of A_μ and $\partial/\partial x_\mu$.

[5]In this section the caret is used to denote all quantities in Euclidean space. The Greek letters $\mu, \nu, \ldots$ denote indices running from 0 to 3 for Minkowskian quantities; for Euclidean quantities (with caret) they run from 1 to 4. The Latin letters $i, j = 1, 2, 3$.

It is quite obvious that for the scalar (Lorentz-invariant) fields ϕ, both real and complex, we can define

$$\phi = \hat{\phi} . \tag{2.24}$$

This concludes the bosonic part of the transition. What remains to be done is deriving similar expressions for the Dirac spinor fields. We begin with the definition of four Hermitean γ-matrices $\hat{\gamma}_\mu$,

$$\begin{gathered} \hat{\gamma}_4 = \gamma_0 , \quad \hat{\gamma}_i = -i\gamma^i , \\ \{\hat{\gamma}_\mu, \hat{\gamma}_\nu\} = 2\delta_{\mu\nu}, \end{gathered} \tag{2.25}$$

where γ_0 and γ^i are the conventional Dirac matrices.

In the Euclidean space the fields ψ and $\bar{\psi}$ over which we integrate in the path integral (see the subsequent lectures) must be regarded as independent anticommuting variables. It is convenient to define the variables $\hat{\psi}$ and $\hat{\bar{\psi}}$ as follows:

$$\psi = \hat{\psi}, \qquad \bar{\psi} = -i\hat{\bar{\psi}} . \tag{2.26}$$

Under rotations in the Minkowski space, $\bar{\psi}$ transforms as $\psi^\dagger \gamma_0$. In the Euclidean space, $\hat{\bar{\psi}}$ transforms as $\hat{\psi}^\dagger$. Indeed, under infinitesimal rotations of the Minkowski space characterized by the parameters $\omega_{\mu\nu}$ the spinor ψ varies as follows:

$$\delta\psi = -\frac{1}{4}\left(\gamma_\mu\gamma_\nu - \gamma_\nu\gamma_\mu\right)\omega^{\mu\nu}\psi . \tag{2.27}$$

One can readily deduce from Eq. (2.27) the variation of $\bar{\psi} = \psi^\dagger\gamma_0$,

$$\begin{aligned} \delta\left(\psi^\dagger \gamma_0\right) &= -\frac{1}{4}\psi^\dagger \gamma_0\gamma_0\left(\gamma_\nu^\dagger\gamma_\mu^\dagger - \gamma_\mu^\dagger\gamma_\nu^\dagger\right)\gamma_0\omega^{\mu\nu} \\ &= \frac{1}{4}\left(\psi^\dagger\gamma_0\right)\left(\gamma_\mu\gamma_\nu - \gamma_\nu\gamma_\mu\right)\omega^{\mu\nu}, \end{aligned} \tag{2.28}$$

so that $\psi_1^\dagger\gamma_0\psi_2$ is a scalar and $\psi_1^\dagger\gamma_0\gamma_\mu\psi_2$ a vector.

In the Euclidean space, the parameters ω_{ij} remain the same as in the Minkowski space, while $\omega_{0j} = i\hat{\omega}_{4j}$. For the variations of $\hat{\psi}$ and $\hat{\psi}^\dagger$ under rotations, we then obtain

$$\delta\hat{\psi} = \frac{1}{4}\left(\hat{\gamma}_\mu\hat{\gamma}_\nu - \hat{\gamma}_\nu\hat{\gamma}_\mu\right)\hat{\omega}_{\mu\nu}\hat{\psi}, \qquad \delta\hat{\psi}^\dagger = -\frac{1}{4}\psi^+\left(\hat{\gamma}_\mu\hat{\gamma}_\nu - \hat{\gamma}_\nu\hat{\gamma}_\mu\right)\hat{\omega}_{\mu\nu}, \tag{2.29}$$

so that $\hat{\psi}_1^\dagger\hat{\psi}_2$ and $\hat{\psi}_1^\dagger\hat{\gamma}_\mu\hat{\psi}_2$ are a scalar and vector, respectively.

Finally, we can write down the Euclidean action of Yang-Mills theory (or QCD with both spinor and scalar quarks),

$$iS = -\hat{S}, \tag{2.30}$$

where

$$\hat{S} = \int d^4\hat{x}\,\left\{\left[\frac{1}{4g^2}\hat{G}^a_{\mu\nu}\hat{G}^a_{\mu\nu} + \hat{\bar{\psi}}\left(-i\hat{\gamma}_\mu\hat{D}_\mu - im\right)\hat{\psi} + i\theta\frac{1}{32\pi^2}\hat{G}^a_{\mu\nu}\hat{\widetilde{G}}^a_{\mu\nu}\right]\right.$$
$$\left. + \hat{D}_\mu\hat{\phi}^\dagger\hat{D}_\mu\hat{\phi} + \tilde{m}^2\hat{\phi}^\dagger\hat{\phi}\right\}. \tag{2.31}$$

It is assumed that $\hat{\psi}$ and $\hat{\phi}$ are columns in the space of flavors (with the flavor and color indices suppressed in (2.31)), and m and $\tilde{m}$ are mass matrices in this space. Note that in the Euclidean space the Levi-Civita tensor $\varepsilon_{\mu\nu\alpha\beta}$ is defined in such a way that $\varepsilon_{1234} = 1$.

The mass matrices can be always chosen to be diagonal.

For completeness I will present in parallel the expression for the Minkowski action in the form

$$S = \int d^4x\,\left\{\left[-\frac{1}{4g^2}G^{a\,\mu\nu}G^a_{\mu\nu} + \bar{\psi}\left(i\gamma^\mu D_\mu - m\right)\psi + \theta\frac{1}{32\pi^2}G^{a\,\mu\nu}\widetilde{G}^a_{\mu\nu}\right]\right.$$
$$\left. + D_\mu\phi^\dagger D^\mu\phi - \tilde{m}^2\phi^\dagger\phi\right\}. \tag{2.32}$$

In what follows we will omit the carets in the Euclidean formalism. Whether we deal with the Minkowski or Euclidean Lagrangian should be clear from the context.

Appendix 2.1

This is the first lecture in this course in which Feynman diagrams are presented. Specific Feynman graphs have their own names. We studied some of them, such as the tree graphs, in the course of QFT I. On page 21 the reader will find other examples.

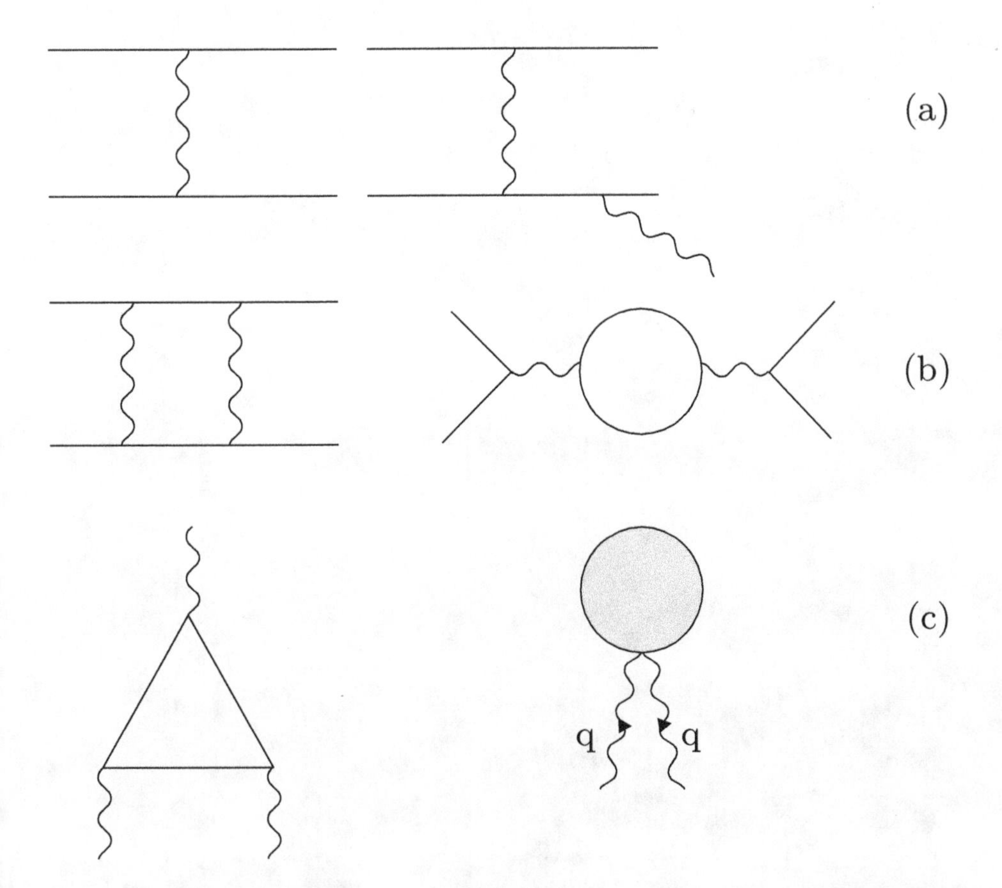

Fig. 2.2 The Feynman diagrams in the line (a) are the tree graphs which we already know from the QFT I course. All other diagrams are loop graphs. On the left-hand side in line (b) you see the so-called box graph while in line (c) the triangular graph. On the right-hand side of line (c) you see the so-called tadpole graph. In the tadpole graph the external line momentum q does not penetrate in the loop.

References

[1] Stefan Pokorski, *Gauge Field Theories*, (Cambridge University Press, 1987).
[2] M. Shifman and A. Vainshtein, Nucl. Phys. B **362**, 21 (1991).

Chapter 3

Lecture 3. Path Integral Quantization

In this lecture the notion of path integral is introduced in the context of quantum mechanics which can be viewed as a one-dimensional field theory. This is the simplest appropriate environment for a pedagogical introduction to the subject.

3.1 Path Integral

Path integral formulation of quantum mechanics and quantum field theory was discovered by Richard P. Feynman who developed and reinterpreted some earlier remarks by Paul Dirac.

3.1.1 *What is path integral*

Assume we have a functional depending on continuous functions. Below we will deal almost exclusively with the action functional. The simplest examples are as follows:

$$S[X(t)] = \int dt\mathcal{L} = \int dt \left[\frac{1}{2}\dot{X}(t)^2 - V(X)\right], \qquad \text{QM}, \quad (3.1)$$

$$S[\phi(t,\vec{x})] = \int dt d\vec{x}\,\mathcal{L}$$

$$= \int dt d\vec{x} \left[\frac{1}{2}\dot{\phi}(t,\vec{x})^2 - \frac{1}{2}m^2\phi^2\right], \quad \text{scalar QFT}, \quad (3.2)$$

where $V(X)$ is a potential in quantum mechanics with one degree of freedom, while (3.2) presents a free field theory for a scalar field ϕ.

Let us first examine (3.1). In this theory the action

$$S = \int dt \left[\frac{1}{2} \dot{X}(t)^2 - V(X) \right] \tag{3.3}$$

is a functional of one real function $X(t)$ with the boundary conditions

$$X(t_1) = X_i \,, \qquad X(t_2) = X_f \tag{3.4}$$

Consider all possible trajectories satisfying (3.4), as shown in Fig. 3.1. If a given trajectory satisfies the classical equation of motion with the boundary conditions (3.4),

$$\left. \frac{\delta S}{\delta X(t)} \right|_{X = X_{\rm cl}} = 0 \tag{3.5}$$

(this is the minimal action trajectory), then we call it the *classical trajectory.* This is the trajectory along which a classical particle moves, see the blue line in Fig. 3.1). All other trajectories are classically inaccessible.

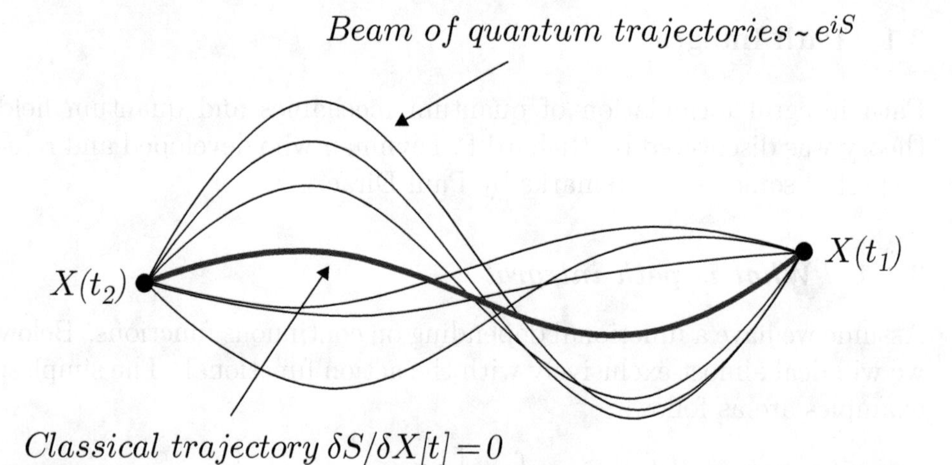

Fig. 3.1 The set of all trajectories connecting $X(t_1) = X_i$ and $X(t_2) = X_f$.

However, a quantum particle "probes" all trajectories, even classically inaccessible. According to Feynman, the probability *amplitude* for a particle constrained by (3.4) roughly speaking can be written as follows. Assume that we discretize a network of trajectories to a countable set,

$$X_{\rm cl}, \; X_{\pm 1}, \; X_{\pm 2}, \; .... \tag{3.6}$$

Then the path integral can be written as follows:

$$\langle X(t_2){=}X_f | X(t_1) = X_i \rangle \equiv \sum_i N \exp\left(iS[X_i(t)]\right)$$

$$\to \int \mathcal{D}X(t) \exp\left(iS[X(t)]\right)_{\text{b.c.}}, \tag{3.7}$$

where N is a normalizing factor which depends on the density of the discretized trajectory network.

Although the conceptual definition is simple, a mathematically pure procedure is technically rather complicated, and so is the determination of N (or, which is the same, of the integration measure). I will not dwell on it, referring the mathematically oriented reader to [1; 2] for details.

Instead, I will present a heuristic argument comparing what we know from the canonic quantization with what follows from the path integral procedure. We will see that both procedures lead to identical results.

The use of the path integral formalism became indispensable in many aspects of modern field theory. Some examples are:

(a) Deriving general properties and relations such as the Ward identities;

(b) Quantization of gauge fields;

(c) Straightforward development of perturbation theory;

(d) Nonperturbative quasiclassical calculations.

(e) Numerical calculations in discretized lattice Euclidean QCD.

The only path integral which can actually be analytically calculated is the Gaussian path integral.

3.1.2 *Quantum-mechanical example: Harmonic oscillator*

For harmonic oscillator

$$V(X) = \frac{1}{2}\omega^2 X^2, \tag{3.8}$$

(see Fig. B.2 on page 254). We will choose $t_1 \to -\infty$, $t_2 \to +\infty$, and $X_i = X_f = 0$. The latter condition guarantees that the classical trajectory is trivial,

$$X(t) = 0, \qquad S(X_{\text{cl}}) = 0. \tag{3.9}$$

We will calculate the energy eigenvalues of the harmonic oscillator using the path integral quantization.

First, it is convenient (although not necessary) to perform a Euclidean rotation. Under this rotation

$$t \to -i\tau\,, dt \to -id\tau \qquad iS_{\rm M} \to i(-i)\int d\tau\left[-\frac{1}{2}\dot{X}(\tau)^2 - \frac{1}{2}\omega^2 X^2\right]$$

$$\to -\int d\tau\left[\frac{1}{2}\dot{X}(\tau)^2 + \frac{1}{2}\omega^2 X^2\right] \equiv -S_{\rm E}\,. \tag{3.10}$$

Note that the Euclidean action $S_{\rm E}$ is positive definite. The weight function $\exp\left(-S_{\rm E}[X(\tau)]\right)$ guarantees the convergence of the path integral.

Now, one can represent an arbitrary trajectory $X(\tau)$ as

$$X(\tau) = X_{\rm cl}(\tau) + \sum_n c_n\, x_n(\tau)\,, \tag{3.11}$$

where $x_n(\tau)$ is a complete set of orthonormal functions that vanish at infinities,

$$\int_{-T/2}^{T/2} d\tau\, x_n(\tau) x_m(\tau) = \delta_{nm}\,, \qquad T \to \infty\,. \tag{3.12}$$

Given that in the case at hand $X_{\rm cl} = 0$, the Euclidean action reduces to

$$S_{\rm E} = \int d\tau \sum_{n,m} (c_m c_n)\, x_n(\tau)\left[-\frac{1}{2}\,\ddot{x}_m(\tau) + \frac{1}{2}\omega^2 x_m(\tau)\right]. \tag{3.13}$$

It is natural to choose the basis functions as the eigenfunctions of the operator below

$$\left[-\frac{d^2}{d\tau^2} + \omega^2\right] x_m(\tau) = \varepsilon_m\, x_m(\tau)\,. \tag{3.14}$$

Since the operator is Hermitean the basis is complete. Combining (3.13) and (3.14) we arrive at

$$S_{\rm E} = \frac{1}{2}\sum_n \varepsilon_n\, c_n^2\,, \tag{3.15}$$

implying, in turn, that

$$\langle X_f|\exp(-HT)\,|X_i\rangle = N\int\left[\prod_n dc_n\,\frac{1}{\sqrt{2\pi}}\right]\exp\left(-\frac{1}{2}\varepsilon_n c_n^2\right) \tag{3.16}$$

where the expression in the square brackets represents $\mathcal{D}X(\tau)$. Indeed, sweeping all values of the coefficients c_n, we cover all possible trajectories.[1] The normalization factors $\frac{1}{\sqrt{2\pi}}$ can be chosen at will, since they combine with the overall normalization N which is not specified so far.[2]

If none of the eigenvalues ε_n vanishes (as is the example under consideration) then $\langle X_f|\exp(-HT)|X_i\rangle$ reduces to[3]

$$\langle X_f|\exp(-HT)|X_i\rangle \to \prod_n \frac{1}{\sqrt{\varepsilon_n}} \stackrel{\text{def}}{=} \left[\det\left(-\frac{d^2}{d\tau^2}+\frac{1}{2}\omega^2\right)\right]^{-1/2}. \tag{3.17}$$

It is customary to set $N = 1$ under the above conventions. The second equality is derived from the theory of ordinary finite-dimensional matrices.

3.2 Path integral quantization

For convenience, I will rewrite Eq. (3.16) as follows

$$\langle X_f = 0|\exp(-HT)|X_i = 0\rangle = N\prod_n \int \left[dc_n \frac{1}{\sqrt{2\pi}} \exp\left(-\frac{1}{2}\varepsilon_n c_n^2\right)\right]. \tag{3.18}$$

The time evolution is indicated explicitly. Equation (3.18) is exact for the harmonic oscillator. Anharmonic terms in the potential, e.g. $\Delta V = X^3 + X^4$ or $\Delta V = X^4 + X^6$ or any other anharmonic terms will give rise to corrections which will ruin factorization of the integrals in (3.18). However, they can be treated in perturbation theory provided that nonlinearity constants are small.

Why are we unable to find the exact path integral if the potential is anharmonic?

The analogs and Eqs. (3.13) and (3.14) will become nonlinear and it will be impossible to diagonalize them. The corresponding matrix

[1] There is a subtlety in the above statement. In the multidimensional space for this statement to be valid one must assume that the space $\{x_1, x_2, ...\}$ is simply connected, i.e. there are no punctured points or domains. Another subtlety – possible vanishing of the energy eigenvalues – will be addressed below.

[2] The $1/\sqrt{2\pi}$ factor is chosen to ensure that

$$\int_{-\infty}^{\infty} \frac{dc}{\sqrt{2\pi}} \exp(-c^2/2) = 1\,.$$

[3] In the Euclidean time the time evolution is given by the operator $\exp(-HT)$. The factor N is implicitly included in the definition of the functional determinant.

will remain infinite-dimensional and highly entangled. There is no way to do such integrals analytically.

Now, let us return to Eq. (3.18). The amplitude (3.18) reduces to

$$\langle X_f = 0 | \exp(-HT) | X_i = 0 \rangle = \prod_n \frac{1}{\sqrt{\varepsilon_n}} = \left[\det \left(-\frac{d^2}{d\tau^2} + \omega^2 \right) \right]^{-1/2},$$

$$\tau_i = -\frac{T}{2}, \quad \tau_f = \frac{T}{2}, \quad T \to \infty . \tag{3.19}$$

The operator $\exp(-HT)$ describes time evolution in the Euclidean time, according to the general rules of quantum mechanics. (In the Minkowski time it would be $\exp(-iHT_{\text{Mink}})$.)

The eigenfunctions of the operator (3.14) are the sine and cosine functions. For instance, the lowest-eigenvalue eigenfunction is

$$\cos\left(\frac{\pi\tau}{T}\right), \tag{3.20}$$

the next eigenfunction is

$$\sin\left(\frac{2\pi\tau}{T}\right), \tag{3.21}$$

the next

$$\cos\left(\frac{3\pi\tau}{T}\right), \tag{3.22}$$

and so on. [Question: why is $x_m(\tau) \equiv 0$ not included?]

Equation (3.14) implies the following set of eigenvalues

$$\varepsilon_n = \frac{\pi^2 n^2}{T^2} + \omega^2, \qquad n = 1, 2, 3, ... \tag{3.23}$$

Next, we split the determinant into two factors,

$$N \left[\det \left(-\frac{d^2}{d\tau^2} + \omega^2 \right) \right]^{-1/2} = N \prod_{n=1}^{\infty} \left(\frac{\pi^2 n^2}{T^2} \right)^{-1/2} \prod_{n=1}^{\infty} \left(1 + \frac{\omega^2 T^2}{\pi^2 n^2} \right)^{-1/2}. \tag{3.24}$$

The factor

$$N \prod_{n=1}^{\infty} \left(\frac{\pi^2 n^2}{T^2} \right)^{-1/2} \tag{3.25}$$

is *omega* independent and hence corresponds to *free* motion of the particle (i.e. setting $\omega = 0$ in Eq. (3.8)). Therefore, it must coincide with the trivial free particle result, namely,

$$\langle X_f = 0| \exp\left(-p^2 T/2\right) |X_i = 0\rangle \equiv \sum_n |\langle p_n|X=0|^2 \, e^{-p_n^2 T/2}$$

$$\longrightarrow \int_{-\infty}^{+\infty} \frac{dp}{2\pi} \exp\left(-\frac{p^2}{2}T\right) = \frac{1}{\sqrt{2\pi T}} \,. \tag{3.26}$$

Of course, Eq. (3.26) is somewhat symbolic: it can be viewed as the definition of N in (3.25).

Now, let us deal with the second factor in (3.24). This infinite product seems monstrous, but – believe it or not – it is well known to mathematicians. Namely,

$$\pi\, y \prod_{n=1}^{\infty} \left(1 + \frac{y^2}{n^2}\right) = \sinh \pi y \,. \tag{3.27}$$

Denoting $y = \omega T/\pi$, we arrive at

$$N \left[\det\left(-\frac{d^2}{d\tau^2} + \omega^2\right)\right]^{-1/2} = \frac{1}{\sqrt{2\pi T}} \left(\frac{\sinh \omega T}{\omega T}\right)^{-1/2}$$

$$\longrightarrow \left(\frac{\omega}{\pi}\right)^{1/2} (2\sinh \omega T)^{-1/2}$$

$$\stackrel{T\to\infty}{\longrightarrow} \left(\frac{\omega}{\pi}\right)^{1/2} \exp(-\omega T/2) \left(1 + \frac{1}{2}\exp(-2\omega T) + \ldots\right). \tag{3.28}$$

Equation (3.28) tells us that for the lowest-energy state $E_0 = \omega/2$ and $|\psi(0)|^2 = (\omega/\pi)^{\frac{1}{2}}$. The next term in the expansion corresponds to the level of the harmonic oscillator with $n = 2$ (the odd values of n do not contribute since for them $\psi(0) = 0$.) Thus, we have

$$E_2 = 5\omega/2 \,, \tag{3.29}$$

and so on. The path integral quantization of the harmonic oscillator produces the level spectrum identical to that we know to emerge in all standard methods.

3.3 Harmonic oscillator: Green's function

In this section, we will return temporarily to the Minkowski space.

Let us see how one can obtain Green's function for the harmonic oscillator using path integral. To this end, let us return to Eq. (3.7) and add a source term in the action,

$$Z[J(\tau)] = \int \mathcal{D}X \exp\left\{ i \int dt \left[\left(\frac{\dot{X}^2}{2} - \frac{\omega^2}{2} X^2 \right) + J(t)X(t) \right] \right\},$$

$$\equiv \int \mathcal{D}X e^{iS_J} \tag{3.30}$$

where $Z[J(\tau)]$ is called the generating functional, because $Z[J]$ can generate all correlation functions through the functional derivatives similar to (3.31). Here $J(\tau)$ is an auxiliary arbitrary source function. It is obvious that by definition [4]

$$Z^{-1} \frac{\delta^2}{\delta J(t)\delta J(t')} Z[J] \bigg|_{J=0} = -\overline{X(t)X(t')}\,, \tag{3.31}$$

where the overline means averaging in the path-integral sense. $Z[J \equiv 0]$ is called the partition function, by analogy with condensed matter. The English term "partition function" comes from German *Zustandsumme*, where *Zustand* means state, hence, the sum over states.

[Question: what are the boundary conditions?]

To save time let us rewrite J-dependent action as follows:

$$S_J = \int dt \left(\frac{1}{2} X D^2 X + JX \right), \tag{3.32}$$

where

$$D^2 \equiv -d^2 - \omega^2\,. \tag{3.33}$$

In Sec. 3.3, we use only the most general properties of the path integrals, such as shifts (3.34), see below. This is the reason why we can, and retain in this section the Minkowski time, rather than passing to the Euclidean time.

Since in (3.30) we have integration $\int \mathcal{D}X$, we can make a shift of variable,

$$X = X' - D^{-2} J \tag{3.34}$$

[4] To be more exact, Eq. (3.31) is valid in the problem at hand since $\left(\overline{X}\right)_{\text{gr. st.}} = 0$. In the general case, it has to be modified in order to eliminate the so-called *disconnected contributions*. The accurate formula is $-\langle \mathrm{T}\{X(t)X(t')\}\rangle = \delta^2 \log Z[J]/\delta J(t)\delta J(t')$ at $J = 0$. For a more detailed explanation see Sec. 4.3.

This shift is t dependent, but the integral is the path integral (i.e. $\mathcal{D}X$ can be replaced by $\mathcal{D}X'$)! Here D^{-2} is the formal inverse operator to D^2. It can be readily defined in the momentum space through the Fourier transformation, see below. I guessed the form of the shift in (3.34) in such a way as to get rid of the $O(X)$ term in the shifted path integral.

Substituting (3.34) in (3.32) we obtain

$$S_J = \int dt\left(\frac{1}{2}X'D^2X' + JX' - \frac{1}{2}X'J - \frac{1}{2}JX' + \frac{1}{2}JD^{-2}J - JD^{-2}J\right)$$

$$= \int dt\left(\frac{1}{2}X'D^2X' - \frac{1}{2}JD^{-2}J\right). \tag{3.35}$$

Please, observe that with the disappearance of the linearity in X term, the linearity in J term disappears too.

Equation (3.35) implies that

$$\overline{X(t)X(t')} = iD^{-2} \longrightarrow i\frac{1}{E^2 - \omega^2}, \tag{3.36}$$

One should remember, of course, that $\omega^2 \to \omega^2 - i\epsilon$.

In the canonic quantization method the same Green's function is defined as *the T-product*

$$\langle 0|T\left(X(t), X(0)\right)|0\rangle$$
$$\equiv \Big\langle 0\Big|\Big(X(t)X(0)\theta(t) + X(0)X(t)\theta(-t)\Big)\Big|0\Big\rangle \tag{3.37}$$

where θ denotes the Heaviside step function and $t' = 0$. The result in (3.36) is obtained through canonic quantization in a rather simple and straightforward way. Namely,

$$D^2\langle 0|T\left(X(t), X(0)\right)|0\rangle = \langle 0|T\left(D^2X(t), X(0)\right)|0\rangle$$
$$-[\dot{X}(t), X(0)]\delta(t) = i\,\delta(t). \tag{3.38}$$

Performing the Fourier transform we arrive at the same formula (3.36). In other words,

$$\overline{X(t)X(t')}\Big|_{\text{path int}} \leftrightarrow \langle 0|T\left(X(t), X(t')\right)|0\rangle\Big|_{\text{canonic quant}}. \tag{3.39}$$

The same is certainly true for all other correlators (n-point functions).

Richard P. Feynman

Appendix 3.1 On Richard Feynman

Richard P. Feynman (1918-1988) was one of the most outstanding theoretical physicist who had shaped modern field theory. His most revolutionary works were the path integral formulation of quantum mechanics and quantum field theory, renormalization of quantum electrodynamics, and the physics of superfluidity of the supercooled liquid helium. In particle physics, he is most known for the breakthrough Feynman-Gell-Mann paper on β decays and for the parton model. He also suggested Feynman diagrams, a universal language of modern theoretical physics.

For his contributions to the development of quantum electrodynamics, Feynman, jointly with Julian Schwinger and Sin-Itiro Tomonaga, received the Nobel Prize in Physics in 1965.

References

[1] Richard P. Feynman and A. R. Hibbs, *Quantum Mechanics and Path Integrals*, (McGraw-Hill, 1965).

[2] Peskin and Schroeder, Chapter 9.

Chapter 4

Lecture 4. Path Integral in Scalar QFT

Partition function a.k.a. state sum. Green's functions and n-point functions. Connected contributions to n-point functions are obtained by functional differentiation of $\log Z[J]$.

4.1 Free scalar field theory

To begin with, we will consider free scalar field theory defined by the Lagrangian

$$L = \frac{1}{2}\partial_\mu\phi(x)\partial^\mu\phi(x) - \frac{m^2}{2}\phi^2(x)\,, \tag{4.1}$$

where ϕ is a *real* field. The partition functional takes the form

$$Z[J] = \int \mathcal{D}\,\phi(x)\exp\left[i\int d^4x\left(L[\phi(x)] + J(x)\phi(x)\right)\right], \tag{4.2}$$

where $J(x)$ is an auxiliary source necessary for determining n-point functions. Sometimes it is convenient to introduce $L_J \equiv L + \Delta L$ where

$$\Delta L = J(x)\,\phi(x)\,. \tag{4.3}$$

Then the partition function (as a functional of $J(x)$) can be rewritten as

$$Z[J] = \int \mathcal{D}\,\phi(x)\exp\left(i\int d^4x\, L_J[\phi(x)]\right). \tag{4.4}$$

4.2 Calculation of the partition function

Keeping in mind that the partition function is determined by the action, not the Lagrangian, we can do integration by parts in the kinetic term;

then

$$L_J \to -\frac{1}{2}\phi(x)\partial^2\phi(x) - \frac{m^2}{2}\phi^2 + J\,\phi$$

$$\equiv \frac{1}{2}\phi(x)\mathcal{P}^2\phi(x) + J\,\phi\,, \tag{4.5}$$

where the operator $\mathcal{P}^2$ is defined as

$$\mathcal{P}^2 = -\partial_\mu\partial^\mu - m^2\,. \tag{4.6}$$

The corresponding action can be written as

$$S_J[\phi(x)] = \int d^4x \left(\frac{1}{2}\phi(x)\mathcal{P}^2\phi(x) + J\,\phi\right). \tag{4.7}$$

Now let us have a closer look at the partition function (4.2) – or, which is the same, state sum – and try to find a shift in the integration variable $\phi(x)$ such that the linear in ϕ term in (4.7) cancels, in much the same way in the harmonic oscillator problem. This is a simple exercise. I will present the final answer and we will check that it does the job.

Indeed, if a new field ϕ' is introduced as

$$\phi(x) = \phi'(x) - \mathcal{P}^{-2}J(x) \tag{4.8}$$

then

$$L_J = \frac{1}{2}\left(\phi' - \mathcal{P}^{-2}J\right)\mathcal{P}^2\left(\phi' - \mathcal{P}^{-2}J\right) + J\left(\phi' - \mathcal{P}^{-2}J\right)$$

$$= \frac{1}{2}\phi'\mathcal{P}^2\phi' + \frac{1}{2}\,J\mathcal{P}^{-2}\,\mathcal{P}^2\,\mathcal{P}^{-2}\,J - 2\times\frac{1}{2}\phi' J + J\,\phi' - J\,\mathcal{P}^{-2}\,J$$

$$= \frac{1}{2}\phi'\mathcal{P}^2\phi' - \frac{1}{2}\,J\mathcal{P}^{-2}\,J\,, \tag{4.9}$$

and, of course,

$$\mathcal{D}\,\phi(x) = \mathcal{D}\,\phi'(x)\,. \tag{4.10}$$

Therefore, in terms of the new field ϕ' the partition function (4.2) takes the form

$$Z[J] = \int \mathcal{D}\,\phi'(x)\exp\left\{i\int d^4x\left[\frac{1}{2}\phi'\mathcal{P}^2\,\phi' - \frac{1}{2}\int d^4x\,d^4y J(x)\mathcal{P}^{-2}J(y)\right]\right\}. \tag{4.11}$$

The second term in the square brackets depends on J but does not depend on ϕ'. Therefore, the J dependence factors out and $Z[J]$ reduces to

$$
\begin{aligned}
Z[J] &= \exp\left(-\frac{i}{2}\int d^4x d^4y J(x)\mathcal{P}^{-2}J(y)\right) \\
&\times \int \mathcal{D}\phi(x)\exp\left(i\int d^4x\,\frac{1}{2}\phi\mathcal{P}^2\,\phi\right) \\
&= Z[J=0]\left(1-\frac{i}{2}\int d^4x\,d^4y J(x)\mathcal{P}^{-2}J(y)+...\right)
\end{aligned}
\tag{4.12}
$$

where ellipses stand for the terms $O(J^4)$, $O(J^6)$, and so on.

4.3 General remark on n-point functions

Equation (4.12) implicitly assumes that the vacuum of the theory on hand is $\phi_{\text{vac}} = 0$. Otherwise from Eq. (4.4) we would conclude that

$$
-i\frac{\{\delta Z[J]/\delta J\}}{Z[J]}\bigg|_{J=0} = \langle\phi\rangle
$$

which should be visible in Eq. (4.12) but it is not! The reason is that the accurate definition of the n-point functions, which includes only the connected contributions is [1]

$$
\frac{\delta^n \log Z[J]}{\delta J(x_1)\,\delta J(x_2)\,...\delta J(x_n)}\bigg|_{J=0}\,. \tag{4.13}
$$

For instance, for $n=2$ we have

$$
\begin{aligned}
\frac{\delta^2 \log Z[J]}{\delta_1\,\delta_2} &= \delta_2\frac{1}{Z}\frac{\delta Z}{\delta_1} \\
&= \frac{1}{Z}\frac{\delta^2 Z}{\delta_1\delta_2} - \frac{1}{Z^2}\frac{\delta Z}{\delta_1}\frac{\delta Z}{\delta_2}
\end{aligned}
\tag{4.14}
$$

where $\delta_k \equiv \delta/\delta J_k$ and setting $J=0$ *after* functional differentiation is not explicitly indicated (but it is implied). The first term in the second line is the two-point function (or, the Green's function) similar to that discussed in Lecture 3. The second term in the second line subtracts $i\langle\phi\rangle i\langle\phi\rangle$. This is the so-called disconnected term because it does not depend on $x_1 - x_2$. In the theory (4.5), the disconnected term vanishes

[1] See footnote on page 30.

since the Lagrangian (4.5) describes the free field theory. To see that disconnected terms do appear if we (functionally) differentiated $Z[J]$ rather than $\log Z[J]$ we must add to (4.5) interaction terms, e.g.

$$\Delta L_{\text{int}} = -\frac{\lambda}{4}\phi^4 . \tag{4.15}$$

In this case, one could have a Higgs-type potential (i.e. $m^2 < 0$, and $\lambda > 0$) resulting in $\phi_{\text{vac}} \equiv \langle\phi\rangle \neq 0$. Then the term $i\langle\phi\rangle i\langle\phi\rangle$ must be subtracted from the connected two-point function, as indicated in Eq. (4.14).

The situation with the disconnected contributions become more pronounced in the four-point functions. For $n = 4$, we have

$$\begin{aligned}
\frac{\delta^4 \log Z[J]}{\delta_1\,\delta_2\,\delta_3\,\delta_4} &= \frac{1}{Z}\delta_1\,\delta_2\,\delta_3\delta_4 Z \\
&- \left(\frac{\delta_1 Z}{Z} \times \frac{\delta_2\delta_3\delta_4 Z}{Z} + \text{permutations}\right) \\
&- \left(\frac{\delta_1\delta_2 Z}{Z} \times \frac{\delta_3\delta_4 Z}{Z} + \text{permutations}\right) \\
&+2\left(\frac{\delta_1 Z}{Z} \times \frac{\delta_2 Z}{Z} \times \frac{\delta_3\delta_4 Z}{Z} + \text{permutations}\right) \\
&-6\left(\frac{\delta_1 Z}{Z} \times \frac{\delta_2 Z}{Z} \times \frac{\delta_3 Z}{Z} \times \frac{\delta_4 Z}{Z}\right),
\end{aligned} \tag{4.16}$$

where $\delta_k \equiv \delta/\delta J_k$. All lines in the expression above other than the first line subtract disconnected contributions. For instance, the product in the third line presents the product of two Green's functions: the first depends on $x_1 - x_2$ while the second on $x_3 - x_4$, see the left-most diagram in Fig. 4.1. Because of permutations, the third line subtracts all three graphs in Fig. 4.1 proportional to

$$[G(x_1,x_2)\,G(x_3,x_4) + G(x_1,x_3)\,G(x_2,x_4) + G(x_1,x_4)\,G(x_2,x_3)] \tag{4.17}$$

This contribution must be subtracted from the connected part. In addition, the second, fourth and fifth lines (they vanish if $\langle\phi\rangle = 0$) subtract those terms that are proportional to $\langle\phi\rangle$.

The disconnected graphs just iterate previously calculated connected graphs.

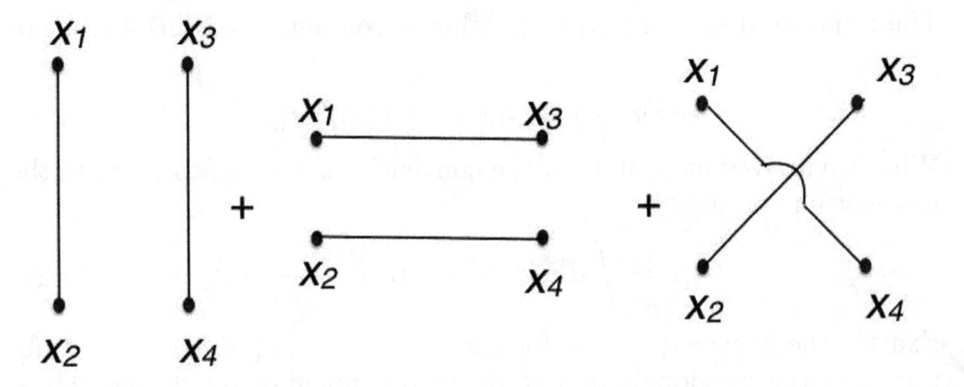

Fig. 4.1 Disconnected Feynman graphs.

Summarizing, the general formula for connected n-point functions is

$$G(x_1, x_2, ..., x_n)_{\text{connected}} = \overline{\phi(x_1)\phi(x_2), ..., \phi(x_n)}$$
$$= (-i)^n \left. \frac{\delta^n \log Z[J]}{\delta J(x_1)\, \delta J(x_2) \,...,\, \delta J(x_n)} \right|_{J=0} . \tag{4.18}$$

Equation (4.18) is our master formula for n-point functions. In what follows, the subscript connected will be omitted.

4.4 Calculation of the scalar propagator

Green's function, or the propagator of the scalar field is by definition given by a (normalized) average of the product $\phi(x)\phi(y)$ over all paths,

$$\langle \phi(x)\phi(y) \rangle = \int \mathcal{D}\,\phi \,\{\phi(x)\phi(y)\}\, e^{i\int d^4x L[\phi(x)]} \tag{4.19}$$

Using Eq. (4.2) or (4.4) we see that (4.19) can be rewritten as[2]

$$\langle \phi(x)\phi(y) \rangle = -\frac{1}{Z[J]} \left. \frac{\delta^2\, Z[J]}{\delta J(x)\, \delta J(y)} \right|_{J=0} . \tag{4.20}$$

[2]Please, compare the master formula (4.18) with Eq. (4.20). We calculate the Green's function in the free field theory, hence, $\delta Z/\delta J = 0$ after taking $J = 0$. Therefore, the second term in the second line of (4.14) vanishes.

The right-hand side immediately follows from our result in (4.12), implying

$$\langle \phi(x)\phi(y)\rangle = G(x,y) \equiv i\,\langle x|\mathcal{P}^{-2}|y\rangle\,. \tag{4.21}$$

What we arrived at is, in fact, the conventional propagator; e.g. in the momentum space,

$$G(p) = \int d^4x\, e^{ipx}\, G(x,0) = \frac{i}{p^2 - m^2}\,. \tag{4.22}$$

Exactly the same expression for the T product $\langle T\{\phi(x)\phi(0)\rangle$ was obtained in the previous semester through canonic quantization. Thus, we see that the path integral method and canonic quantization are equivalent.

Chapter 5

Lecture 5. Complex Scalar Field and Finite Temperature

Truncation and vertices. Partition function at finite temperature. What are the appropriate boundary conditions?

5.1 Scalar field (*continued*)

We will make a miniscule step forward compared to the previous lecture if (instead of a real field) we will deal with complex fields $\phi(x)$ and $\phi^\dagger(x)$. Then the Lagrangian L_J with the source terms added takes the form

$$\begin{aligned}\mathcal{L}_J &= -\phi^\dagger(x)\partial^2\phi(x) - m^2\phi^\dagger(x)\phi(x) + J(x)\phi(x) + J^\dagger(x)\phi^\dagger(x) \\ &\equiv \phi^\dagger(x)\mathcal{P}^2\phi(x) + J(x)\phi(x) + J^\dagger(x)\phi^\dagger(x)\,, \qquad (5.1)\end{aligned}$$

where $\mathcal{P}^2$ is defined in Eq. (4.6) as previously. The state sum is

$$Z_J \equiv \int \mathcal{D}\phi(x)\exp\left(i\int d^4x\,\mathcal{L}_J[\phi(x)]\right). \qquad (5.2)$$

The J dependence in (5.2) can be determined by performing a shift of variables in the path integral,

$$\begin{aligned}\phi(x) &= \phi'(x) - \mathcal{P}^{-2}J^\dagger(x)\,, \\ \phi^\dagger(x) &= \phi'^\dagger(x) - \mathcal{P}^{-2}J(x) \qquad (5.3)\end{aligned}$$

where $\phi'(x)$ and $\phi'^\dagger(x)$ are new integration variables,

$$\mathcal{D}\phi(x) = \mathcal{D}\phi'(x)\,, \quad \mathcal{D}\phi^\dagger(x) = \mathcal{D}\phi'^\dagger(x)$$

and $\mathcal{P}^{-2}$ is the inverse operator $\mathcal{P}^2$. It is written in a symbolic form, so that

$$\mathcal{P}^{-2}\mathcal{P}^2 = 1\,. \qquad (5.4)$$

The equation above is a shorthand. It would be mathematically correct to write it as follows

$$\int d^4y\, \mathcal{P}^{-2}(x,y)\mathcal{P}^2(y,z) = \delta(x-z)\,. \tag{5.5}$$

I will use shorthand whenever there is no menace of confusion. Fully explicit expression for $\mathcal{P}^{-2}$ will be given shortly (in the momentum space). It is the same as in the previous Lecture.

Substituting (5.3) in (5.2) we arrive at

$$\begin{aligned} Z_J &\equiv \int \mathcal{D}\phi'(x) \exp\left(i\int d^4x\, \mathcal{L}_{J=0}\right) \\ &\quad \times \exp\left(-i\int d^4x\, d^4y\, J^\dagger(x)\mathcal{P}^{-2}(x,y)J(y)\right) \\ &= Z_{J=0} \times \exp\left(-i\int d^4x\, d^4y\, J^\dagger(x)\mathcal{P}^{-2}(x,y)J(y)\right), \end{aligned} \tag{5.6}$$

cf. (4.12). The J dependence is completely factored out.

Furthermore, n-point Green's functions $G(x_1, x_2, ..., x_n)$ are obtained by applying functional derivatives over $J(x_i)$ or $J^\dagger(y_i)$ n times. The functional derivative acts as follows:

$$\frac{\delta}{\delta J(x)}\, J(y) = \delta(x-y)\,, \tag{5.7}$$

implying that

$$\frac{\delta}{\delta J(x)} \int d^4y J(y)F(y) = F(x) \tag{5.8}$$

for any function F.

If the Lagrangian L of the theory under consideration has a global U(1) symmetry, then the number of functional derivatives over J must be equal to that over $J^\dagger$; otherwise, the result will automatically vanish.

If we do not specify the boundary conditions (the only assumption is that $t_f - t_i = T \to \infty$) then in the Euclidean space we have

$$\langle\phi_f|e^{-TH_J}|\phi_i\rangle = \sum_{m,k} \langle\phi_f|m\rangle\,\langle m|e^{-TH_J}|k\rangle\,\langle k|\phi_i\rangle \tag{5.9}$$

where m, k are energy eigenvalues. Since

$$\langle m|e^{-TH_J}|k\rangle = \delta_{m,k}\exp(-TE_k)$$

at $T \to \infty$ only the ground state survives in the sum.

Hence, the n-point Green's functions $G(x_1, x_2, ..., x_n)$ are the vacuum expectation values

$$G(x_1, x_2, ..., x_n) = \langle 0|\phi(x_1), \dots, \phi^\dagger(x_n)|0\rangle$$

$$= \frac{(-i)^n}{Z_{J=0}} \left[\frac{\delta}{\delta J(x_1)} \cdots \frac{\delta}{\delta J^\dagger(x_n)} Z_J\right]_{J=0} - \text{disconnected contributions}\,, \tag{5.10}$$

(see Sec. 4.3 on disconnected contributions).

Note that the T product appearing in the canonic quantization method is automatic in the path integral quantization.

So far we dealt with a symbolic expression for $\mathcal{P}^{-2}$ in the coordinate space. Now we can write down the actual expression in the momentum space.

The operator $\mathcal{P}^2$ is defined in Eq. (5.1),

$$\left(-\partial^2 - m^2\right)\phi(x) \equiv \mathcal{P}^2\phi(x)\,. \tag{5.11}$$

In the momentum space

$$\left(p^2 - m^2\right)\phi(p) \equiv \mathcal{P}^2\phi(p)\,. \tag{5.12}$$

Thus, in the momentum space the operator $\mathcal{P}^2$ acts multiplicatively, and its inversion is trivial (and local),

$$\mathcal{P}^{-2} = \frac{1}{p^2 - m^2}\,. \tag{5.13}$$

Equation (5.6) results in the following two-point (Green's) function:[1]

$$G(p) = i\mathcal{P}^{-2} \to \frac{i}{p^2 - m^2}\,. \tag{5.14}$$

We had obtained this expression previously (QFT I) using the canonic quantization method.

*****

In order to obtain a non-vanishing (connected) contribution in

$$\left[\frac{\delta}{\delta J(x_1)} \cdots \frac{\delta}{\delta J^\dagger(x_4)} (-i)^4 \log Z_J\right]_{J=0} \tag{5.15}$$

one must add to the Lagrangian (5.1) an interaction term, for instance

$$\mathcal{L}_{\text{int}} = -\frac{\lambda}{4}\left(\phi^\dagger(x)\phi(x)\right)^2\,. \tag{5.16}$$

[1]The functional differentiation to be performed is $\delta^2/\delta J^\dagger(x)\delta J(y)$. Please, explain why $\delta^2/\delta J(x)\delta J(y)$ makes no sense in the case at hand.

Of course, with the quartic term included in the Lagrangian the path integral becomes non-Gaussian, and therefore incalculable. However, if $\lambda \ll 1$, we can expand in λ and deal with the functional integral

$$\int \mathcal{D}\phi(x)\, \frac{-i\lambda}{4} \int d^4y\, \left(\phi^\dagger(y)\phi(y)\right)^2 \exp\left(i \int d^4x\, \mathcal{L}_J[\phi(x)]\right) \tag{5.17}$$

(to the leading order in λ). The first factor in (5.17) is $S_{\rm int}$. The amplitude which is of interest for us can be written as follows:

$$\begin{aligned} i\,\mathcal{A}_{\rm connected} = -i\, i^4\, \lambda \int & d^4y\, d^4x_1\, d^4x_2\, d^4x_3\, d^4x_4 \\ & \times G(y,x_1) J^\dagger(x_1) G(y,x_2) J^\dagger(x_2) \\ & \times G(x_3,y) J(x_3) G(x_4,y) J(x_4). \end{aligned} \tag{5.18}$$

see Fig. 5.1 on page 43. Please verify that all numerical factors (i.e. $1/4$ in front of λ in (5.17) and $1/4!$ from the expansion of the exponent) cancel because of combinatorics.

Amputating (or, which is the same as truncating) the legs represented by the factors $G(y, x_1)$, etc., we deduce that the vertex is

$$\mathcal{A}_{\rm connected} = -\lambda\,. \tag{5.19}$$

5.2 QFT at finite temperature

Consider the same theory in Euclidean. Then

$$\langle \phi(\vec{x}, \tau_f)|\exp(-H\mathcal{T})|\phi(\vec{x},\tau_i)\rangle = \int_{\phi(\tau_i)=\phi_i(\vec{x})}^{\phi(\tau_f)=\phi_f(\vec{x})} \mathcal{D}\phi(x)\exp(-S) \tag{5.20}$$

where $\mathcal{T} = \tau_f - \tau_i$. To make contact with the thermal field theory it is convenient to denote

$$\mathcal{T} = \beta = \frac{1}{T} \tag{5.21}$$

where T stands for *temperature* in what follows.

Now, we impose the additional condition

$$\phi_i(\vec{x}) = \phi_f(\vec{x}) \tag{5.22}$$

and sum (integrate) over all $\phi_i(\vec{x}) = \phi_f(\vec{x}) \equiv \phi_{\rm p.b.c.}$ (periodic boundary conditions),

$$\sum_{\phi_{\rm p.b.c.}(\vec{x})} \langle \phi_{\rm p.b.c.}(\vec{x})|\exp(-H\mathcal{T})|\phi_{\rm p.b.c.}(\vec{x})\rangle = \int_{\phi(\tau_i)=\phi(\vec{x})}^{\phi(\tau_i+\beta)=\phi(\vec{x})} \mathcal{D}\phi(x)\exp(-S) \tag{5.23}$$

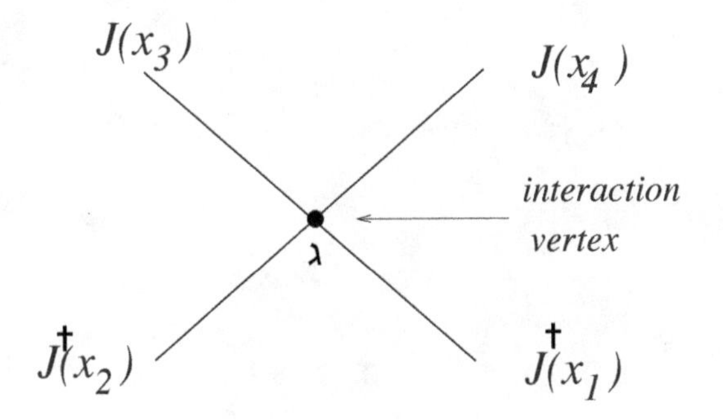

Fig. 5.1 $\left(\phi^\dagger(x)\phi(x)\right)^2$ vertex. See page 42.

with arbitrary $\phi(x)$. Now we path-integrate over all fields periodic (in Euclidean) with the period β. Effectively, we converted the plane four-dimensional space R^4 into a "cylinder" $R^3 \times S^1$. The circumference of the compactified time is β.

Since the bra and ket states on the left-hand side of (5.23) form a complete state, the left-hand side is the trace over the Hilbert space, so we find the relation

$$\operatorname{Tr}\exp(-H/T) = \int_{\phi(\tau_i)=\phi(\vec{x})}^{\phi(\tau_i+\beta)=\phi(\vec{x})} \mathcal{D}\phi(x)\exp(-S)\,. \tag{5.24}$$

The left-hand side of (5.24) is the thermal partition function of a system with Hamiltonian H,

$$Z_T = \operatorname{Tr}\exp(-H/T)\,. \tag{5.25}$$

Sometimes the subscript T is omitted. Then one should understand from the context that it is the thermal partition function that is under consideration.

Equation (5.24) means that the thermal average of the system can be computed using the path integral restricted to paths periodic in Euclidean time (the Matsubara[2] formula, or the Matsubara boundary conditions). The period is the inverse temperature.

Fermions at finite temperature are treated in Sec. 7.2, page 64.

[2]Takeo Matsubara (1921-2014) was a Japanese physicist who proposed a method of treatment of statistical mechanics by applying quantum field theory techniques to statistical physics (1955). He was the winner of the Nishina Memorial Prize in 1961, and later became the director of the Physical Society of Japan.

Chapter 6

Lecture 6. Calculus of Grassmann Numbers and Spinor Fields

Spinors in four, three, and two dimensions. Dirac, Weyl and Majorana spinors.

6.1 Grassmann numbers

First we will have to discuss the so-called Grassmann numbers, a mathematical construction which allows a path integral representation for Fermionic fields.

The Grassmann numbers are almost the same as "regular" numbers Out of all basic rules applicable to conventional numbers (also known as c-numbers) only one is changed, namely the rule of commutativity. The Grassmann numbers anticommute. If x, y are c-numbers, which commute when multiplied, and θ_1, θ_2 are Grassmann numbers, which anticommute, then

$$xy = yx \quad \text{(regular numbers commute)}, \tag{6.1}$$

and

$$\theta_1\theta_2 = -\theta_2\theta_1 \quad \text{(Grassmann numbers anticommute)}. \tag{6.2}$$

Grassmann numbers and c-numbers commute,

$$x\theta_1 = \theta_1 x \quad \text{(c-numbers commute with Grassmann numbers)}. \tag{6.3}$$

If we have a single Grassmann variable θ then, because of the anticommutativity,

$$\theta^2 = 0\,. \tag{6.4}$$

The property of associativity is preserved:

$$(\theta_1\theta_2)\theta_3 = \theta_1(\theta_2\theta_3)\,. \tag{6.5}$$

Grassmann number-based calculus (differentiation and integration) was developed in the 1960s by a Soviet mathematician Felix Berezin [1], see page 57. We will need to use Grassmann differentiation and integration (the latter is usually referred to as the Berezin integral).

6.2 Calculus on Grassmann numbers

Assume that we have a single Grassmann variable θ. Then the Taylor expansion of any function $f(\theta)$ reduces to a binomial,

$$f(\theta) = a + \beta\theta\,. \tag{6.6}$$

If f is a c-numerical function, then a is a c-number while β is a Grassmann number, see Eq. (6.4).

The rules of differentiation are most simple,

$$\frac{\partial a}{\partial\theta} = 0\,, \qquad \frac{\partial}{\partial\theta}(\beta\theta) = -\beta\frac{\partial}{\partial\theta}\theta = -\beta\,. \tag{6.7}$$

If we differentiate a term containing several factors (both c-numbers and Grassmann numbers) over θ, each time we drag ∂_θ to the right through a Grassmann number we change the sign, see Eq. (6.2). Thus $\partial_\theta \equiv \partial/\partial\theta$ has the same Grassmann nature as θ. Basically we have to use only one differentiation rule:

$$\frac{\partial\theta_i}{\partial\theta_j} = \delta_{ij}\,. \tag{6.8}$$

Now let us pass to Grassmann integration. For c-numeric integrals we obviously have

$$\int_{-\infty}^{\infty} dx\, f(x+a) = \int_{-\infty}^{\infty} dx\, f(x)\,, \tag{6.9}$$

provided the integral is convergent. This is the shift property. Let us try to keep it for the functions (6.6) in the Berezin integral, namely,

$$\int d\theta\, f(\theta+\eta) = \int d\theta\, f(\theta)\,. \tag{6.10}$$

Equation (6.6) implies

$$\int d\theta\,[a + \beta(\theta+\eta)] = \int d\theta(a+\beta\theta)\,, \tag{6.11}$$

which in turn leads to the condition

$$\int d\theta\,(\beta\eta) = 0\,. \tag{6.12}$$

Observing that $\beta\eta$ acts as a c-number (being a product of two Grassmann numbers) we conclude that the integral

$$\int d\theta\, a = 0 \tag{6.13}$$

must vanish, while the integral $\int d\theta\,\theta$ can be defined as

$$\int d\theta\,\theta = 1\,. \tag{6.14}$$

Indeed, $d\theta\,\theta$ acts as a c-number. The unity on the right-hand side of (6.14) is a convenient choice of normalization suggested by Berezin.

The multiple Berezin integral is treated in the manner of consecutive integrations, e.g.

$$\int d\theta d\eta(\theta\eta) = -\int d\theta\theta \int d\eta\eta = -1\,. \tag{6.15}$$

For convergent c-numeric integrals, we obviously have

$$\int_{-\infty}^{\infty} d(cx)\, f(c\,x) = \int_{-\infty}^{\infty} dx\, f(x)\,, \tag{6.16}$$

where c is a c-number. Requiring the same from the Berezin integrals and using Eq. (6.6), we conclude that

$$d(c\,\theta) = \frac{1}{c}\, d\theta\,. \tag{6.17}$$

Note that

$$\frac{\partial}{\partial\theta} = \int d\theta\,, \tag{6.18}$$

and the role of the Dirac δ-function is played by θ itself,

$$\delta(\theta) = \theta\,. \tag{6.19}$$

6.3 Generalization to complex Grassmann numbers

In most cases in quantum field theory of fermions we deal with pairs of complex Grassmann numbers θ and $\bar{\theta}$. If $\eta_{1,2}$ are real Grassmann numbers of Sec. 6.2 one may think of θ, $\bar{\theta}$ as

$$\theta = \eta_1 + i\eta_2\,, \qquad \bar{\theta} = \eta_1 - i\eta_2\,. \tag{6.20}$$

The Berezin integral is now defined as

$$\int d\theta\, d\bar{\theta}\,\bar{\theta}\theta = 1\,, \quad \int d\theta\, d\bar{\theta}\bar{\theta} = 0,$$

$$\int d\theta\, d\bar{\theta}\theta = 0, \quad \int d\theta\, d\bar{\theta} = 0\,. \tag{6.21}$$

Everything else remains intact, e.g.

$$\frac{\partial\theta}{\partial\theta} = 1\,, \quad \frac{\partial\bar{\theta}}{\partial\theta} = 0\,, \quad \frac{\partial\bar{\theta}}{\partial\theta} = 0\,, \tag{6.22}$$

etc.

It is convenient to define conjugation as

$$\overline{\theta\eta} = \bar{\eta}\,\bar{\theta}\,. \tag{6.23}$$

6.4 Exponential integrand

Finally, in what follows we will need the integral of a special type over multiple Grassmann variables, θ_1, $\bar{\theta}_1$, θ_2, $\bar{\theta}_2$..., namely,

$$I_N \equiv \int \left(\prod_{n=1}^{N} d\bar{\theta}_n d\theta_n \right) \exp\left(-\bar{\theta}_i A^{ij} \theta_j\right), \tag{6.24}$$

where A^{ij} is an arbitrary *c*-numeric $N \times N$ matrix. This integral can be calculated in the general form using the rules formulated in the previous subsection. Let us first do simple examples corresponding to $N = 1, 2$ and then generalize them.

For $N = 1$, we have

$$I_1 = \int d\bar{\theta}_1 d\theta_1 \exp\left(-\bar{\theta}_1 A^{11} \theta_1\right) = \int d\bar{\theta}_1 d\theta_1 \left(-\bar{\theta}_1 A^{11} \theta_1\right) = A_{11}\,. \tag{6.25}$$

For $N = 2$, we have

$$\begin{aligned} I_2 &= \int d\bar{\theta}_1 d\theta_1 d\bar{\theta}_2 d\theta_2 \, \frac{1}{2} \left(\sum_{i,j=1,2} -\bar{\theta}_i A^{ij} \theta_j \right)^2 \\ &= \int d\bar{\theta}_1 d\theta_1 d\bar{\theta}_2 d\theta_2 \, \frac{1}{2} \left(\bar{\theta}_1 A^{11} \theta_1 + \bar{\theta}_1 A^{12} \theta_2 + \bar{\theta}_2 A^{21} \theta_1 + \bar{\theta}_2 A^{22} \theta_2\right)^2 \\ &= \int d\bar{\theta}_1 d\theta_1 d\bar{\theta}_2 d\theta_2 \left(\bar{\theta}_1 A^{11} \theta_1 \, \bar{\theta}_2 A^{22} \theta_2 + \bar{\theta}_1 A^{12} \theta_2 \, \bar{\theta}_2 A^{21} \theta_1\right) \\ &= A^{11} A^{22} - A^{12} A^{21} = \det A\,. \end{aligned} \tag{6.26}$$

Continuing in this way it is not difficult to see that for any N

$$I_N = \det A\,. \tag{6.27}$$

I leave this as a warm-up exercise before you delve into your home assignment.

6.5 Four-dimensional spinors

Let us start from four dimensions (see e.g. [2]). Four-dimensional spinors realize irreducible representation of the Lorentz group which has six generators: three spatial rotations and three Lorentz boosts.

There are two types of spinors, right-handed and left-handed, which are marked by dotted and undotted indices, respectively, as follows:[1]

$$\text{right-handed:} \quad \bar{\eta}^{\dot{\alpha}}\,, \quad \dot{\alpha} = 1, 2\,, \tag{6.28}$$

$$\text{left-handed:} \quad \xi_\alpha\,, \quad \alpha = 1, 2\,. \tag{6.29}$$

Let us write the transformation law of the undotted spinors as

$$\tilde{\xi}_\alpha = U_\alpha^{\ \beta}\,\xi_\beta\,, \tag{6.30}$$

where for spatial rotations

$$U_{\rm rot} = \exp\left(i\frac{\theta}{2}\vec{n}\vec{\sigma}\right), \quad \det U = 1\,, \quad U^\dagger U = 1\,, \tag{6.31}$$

θ is the rotation angle and $\vec{n}$ is the rotation axis, while for the Lorentz boosts

$$U_{\rm boost} = \exp\left(-\frac{\phi}{2}\vec{n}'\vec{\sigma}\right), \quad \det U = 1\,, \quad U^\dagger U \neq 1\,. \tag{6.32}$$

Here $\tanh\phi = v$, and v is the three-velocity, while $\vec{n}'$ is its direction. Moreover, $\vec{\sigma}$ are the Pauli matrices. As is seen from Eq. (6.32), the transformation matrix for the Lorentz boost is not unitary. This is due to the fact that the Lorentz group is noncompact $O(1,3)$ rather than the compact $O(4)$. If we passed from the Minkowski to Euclidean space the Euclidean "Lorentz group" would be O(4). By definition

$$\bar{\eta}_{\dot{\alpha}} \sim (\xi_\alpha)^*, \tag{6.33}$$

where the sign $\sim$ means "is transformed as". Therefore, for the dotted spinors the Lorentz transformation goes with the complex conjugated matrix[2]

$$\tilde{\bar{\eta}}_{\dot{\alpha}} = (U^*)_{\dot{\alpha}}^{\ \dot{\beta}}\,\bar{\eta}_{\dot{\beta}} \quad \text{or} \quad \tilde{\bar{\eta}}^{\dot{\alpha}} = \begin{cases} (U_{\rm rot})_{\dot{\beta}}^{\ \dot{\alpha}}\,\bar{\eta}^{\dot{\beta}}\,, & \text{for rotations}\,, \\ \left(U_{\rm boost}^{-1}\right)_{\dot{\beta}}^{\ \dot{\alpha}}\,\bar{\eta}^{\dot{\beta}}\,, & \text{for boosts}\,. \end{cases} \tag{6.34}$$

---

[1]This convention is standard in supersymmetry but is opposite to the one accepted in the textbook [2] where the left-handed spinor is dotted. Sometimes we will omit spinorial indices. Then, in order to differentiate between the left- and right-handed spinors, we will mark the latter by bars, e.g. $\bar{\eta}$ is a shorthand for $\bar{\eta}^{\dot{\alpha}}$.

[2]Raising and lowering the spinorial indices is performed by virtue of the Levi-Civita tensors, see Eqs. (6.36) and (6.37) and below. For the derivation of (6.34) from (6.31) see Appendix 6.2.

If three generators of the spatial rotations are denoted by L^i and three Lorentz boost generators by N^i, it is obvious[3] that $L^i + iN^i$ do not act on ξ_α while $L^i - iN^i$ do not act on $\bar{\eta}^{\dot{\alpha}}$ $(i = 1, 2, 3)$. The spinors ξ_α and $\bar{\eta}^{\dot{\alpha}}$ are referred to as the *chiral* or *Weyl* spinors.[4]

In four dimensions one chiral spinor is equivalent to one Majorana spinor, while two chiral spinors – one dotted and one undotted – comprise one Dirac spinor (see below).

In order to be invariant, every spinor equation must have on each side the same number of undotted and dotted indices, otherwise the equation becomes invalid under the change of the reference frame. We must remember, however, that taking the complex conjugate implies interchanging dotted and undotted indices. For instance, the relation $\bar{\eta}^{\dot{\alpha}\dot{\beta}} = \left(\xi^{\alpha\beta}\right)^*$ is invariant.

To build Lorentz scalars we must convolute either undotted or dotted spinors (separately). For instance, the products

$$\chi^\alpha \xi_\alpha \text{ or } \bar{\psi}_{\dot{\beta}} \bar{\eta}^{\dot{\beta}} \tag{6.35}$$

are invariant under the Lorentz transformations. Lowering and raising of the spinorial indices is done by applying the invariant Levi-Civita tensor from *the left*,[5]

$$\chi^\alpha = \varepsilon^{\alpha\beta}\chi_\beta\,, \quad \chi_\alpha = \varepsilon_{\alpha\beta}\chi^\beta, \tag{6.36}$$

and the same for the dotted indices. The two-index Lorentz-invariant Levi-Civita tensor is defined as[6]

$$\begin{aligned} \varepsilon^{\alpha\beta} &= -\varepsilon^{\beta\alpha}\,, \quad \varepsilon^{12} = -\,\varepsilon_{12} = 1\,, \\ \varepsilon^{\dot{\alpha}\dot{\beta}} &= -\varepsilon^{\dot{\beta}\dot{\alpha}}\,, \quad \varepsilon^{\dot{1}\dot{2}} = -\,\varepsilon_{\dot{1}\dot{2}} = 1\,. \end{aligned} \tag{6.37}$$

We will follow a standard shorthand notation

$$\eta\chi \equiv \eta^\alpha\chi_\alpha\,, \qquad \bar{\eta}\bar{\chi} \equiv \bar{\eta}_{\dot{\alpha}}\bar{\chi}^{\dot{\alpha}}\,. \tag{6.38}$$

[3]On the left-handed states $\vec{L} = \frac{1}{2}\vec{\sigma}$, $\vec{N} = \frac{i}{2}\vec{\sigma}$. The algebra of these generators is as follows: $[L^i, L^j] = i\varepsilon^{ijk}L^k$, $[L^i, N^j] = i\varepsilon^{ijk}N^k$ and $[N^i, N^j] = -i\varepsilon^{ijk}L^k$, implying that $[L^i - iN^i\,, L^j + iN^j] = 0$. Note that under spatial rotations ξ_α and $\bar{\eta}^{\dot{\alpha}}$ transform in the same way. This is not the case for Lorentz boosts.

[4]Hermann Weyl (1885-1955) was an outstanding German mathematician and mathematical physicist.

[5]The very same rule, multiplying by the Levi-Civita tensor from the left, applies to quantities with several spinorial indices, dotted, undotted or mixed.

[6]Equation (6.37) implies the following formula: $\varepsilon_{\alpha\gamma}\varepsilon^{\gamma\beta} = \delta_\alpha^\beta$ which is natural. Its consequence is $\varepsilon_{\alpha\gamma}\varepsilon_{\beta\tilde{\gamma}}\,\varepsilon^{\gamma\tilde{\gamma}} = \varepsilon_{\beta\alpha}$.

Please, note that this convention acts differently for the left and right-handed spinors. This convention is very convenient because

$$(\eta\chi)^\dagger = (\eta^\alpha\chi_\alpha)^\dagger = (\chi_\alpha)^*(\eta^\alpha)^* = \bar{\chi}\bar{\eta}\,, \tag{6.39}$$

where

$$\bar{\chi}_{\dot\alpha} \equiv (\chi_\alpha)^*\,, \qquad \bar{\eta}^{\dot\alpha} \equiv (\eta^\alpha)^*\,. \tag{6.40}$$

Moreover, using the properties (6.37) of the Levi–Civita tensor, and the Grassmannian nature of the fermion variables, we get

$$\chi^\alpha\chi^\beta = -\frac{1}{2}\varepsilon^{\alpha\beta}\chi^2\,, \quad \chi_\alpha\chi_\beta = \frac{1}{2}\varepsilon_{\alpha\beta}\chi^2\,,$$

$$\bar{\chi}^{\dot\alpha}\bar{\chi}^{\dot\beta} = \frac{1}{2}\varepsilon^{\dot\alpha\dot\beta}\bar{\chi}^2\,, \quad \bar{\chi}_{\dot\alpha}\bar{\chi}_{\dot\beta} = -\frac{1}{2}\varepsilon_{\dot\alpha\dot\beta}\bar{\chi}^2\,. \tag{6.41}$$

In addition to Lorentz scalars (6.38) one can build a Lorentz four-vector from two spinors, namely

$$v^\mu = \eta^\alpha\,(\sigma^\mu)_{\alpha\dot\beta}\,\bar{\chi}^{\dot\beta}\,, \tag{6.42}$$

where $(\sigma^\mu)_{\alpha\dot\beta}$ represents a set of four 2×2 matrices,

$$(\sigma^\mu)_{\alpha\dot\beta} = \{1, \vec{\sigma}\}_{\alpha\dot\beta} \tag{6.43}$$

Under the Lorentz group v^μ transforms as $\left\{\frac{1}{2},\,\frac{1}{2}\right\}$. Other representations can be constructed from two spinors as follows:

$$\eta_{\{\alpha}\chi_{\beta\}}\,, \quad \bar{\eta}_{\{\dot\alpha}\bar{\chi}_{\dot\beta\}} \tag{6.44}$$

where the braces denote index symmetrization. The representations (6.44) are $\{1,\,0\}$ and $\{0,\,1\}$. Each has three elements.

If the matrix $(\sigma^\mu)_{\alpha\dot\beta}$ is "right-handed", it is convenient to introduce its "left-handed" counterpart,

$$(\,\bar{\sigma}^\mu\,)^{\dot\beta\alpha} = \{\,1\,, -\vec{\sigma}\,\}_{\dot\beta\alpha}\,. \tag{6.45}$$

To obtain the matrix $(\bar{\sigma}^\mu)^{\dot\beta\alpha}$ from $(\sigma^\mu)_{\alpha\dot\beta}$, we raise the indices of the latter according to the rule (6.36) and then transpose the dotted and undotted indices. It is worth remembering that

$$(\,\sigma^\mu\,)_{\alpha\dot\beta}\,(\,\bar{\sigma}^\nu\,)^{\dot\beta\gamma} + (\,\sigma^\nu\,)_{\alpha\dot\beta}\,(\,\bar{\sigma}^\mu\,)^{\dot\beta\gamma} = 2\,g^{\mu\nu}\,\delta^\gamma_\alpha\,. \tag{6.46}$$

An immediate consequence is

$$(\,\sigma^\mu\,)_{\alpha\dot\beta}\,(\,\sigma^\nu\,)_\gamma^{\ \dot\beta} + (\,\sigma^\nu\,)_{\alpha\dot\beta}\,(\,\sigma^\mu\,)_\gamma^{\ \dot\beta} = -2\,g^{\mu\nu}\,\varepsilon_{\alpha\gamma}\,. \tag{6.47}$$

Yet another consequence is

$$(\sigma^\mu)_{\gamma\dot\delta}\,(\bar{\sigma}_\mu)^{\dot\beta\alpha} = 2\,\delta^{\dot\beta}_{\dot\delta}\,\delta^\alpha_\gamma\,. \tag{6.48}$$

We pause here to define two other matrices which are useful in discussing the transformation laws of the Weyl spinors with respect to the Lorentz rotations. One can combine (6.31) and (6.32) in a unified formula if one introduces

$$\sigma^{\mu\nu} \equiv \frac{1}{4}\left(\sigma^\mu\bar{\sigma}^\nu - \sigma^\nu\bar{\sigma}^\mu\right) = \left(-\frac{1}{2}\vec{\sigma}\,,\ \frac{i}{2}\vec{\sigma}\right),$$

$$\bar{\sigma}^{\mu\nu} \equiv \frac{1}{4}\left(\bar{\sigma}^\mu\sigma^\nu - \bar{\sigma}^\nu\sigma^\mu\right) = \left(\frac{1}{2}\vec{\sigma}\,,\ \frac{i}{2}\vec{\sigma}\right). \tag{6.49}$$

In Eq. (6.49), I use a standard shorthand for two-index antisymmetric tensor, namely,

$$\sigma^{\mu\nu} \leftrightarrow \left(\vec{a},\vec{b}\right)$$

means that

$$\sigma^{\mu\nu} = \begin{pmatrix} 0 & a_1 & a_2 & a_3 \\ a_1 & 0 & -b_3 & b_2 \\ a_2 & b_3 & 0 & -b_1 \\ a_3 & -b_2 & b_1 & 0 \end{pmatrix} \tag{6.50}$$

Note that $\sigma^{\mu\nu}$ must act on the left-handed spinors with the lower indices, while $\bar{\sigma}^{\mu\nu}$ on the right-handed spinors with the upper indices.

Let us now return to the issue of constructing the Dirac and Majorana spinors from the Weyl spinors. The Dirac spinors, also known as bispinors, naturally appear in vector-like theories such as Quantum Chromodynamics (QCD). They can be obtained as follows:

$$\Psi = \begin{pmatrix} \xi_\alpha \\ \bar{\eta}^{\dot{\alpha}} \end{pmatrix}. \tag{6.51}$$

Each Dirac spinor requires one left- and one right-handed Weyl spinor. Sometimes, instead of (6.51) the following notation is used

$$\Psi \equiv \Psi_L + \Psi_R\,, \qquad \Psi_L = \begin{pmatrix} \xi_\alpha \\ 0 \end{pmatrix}, \quad \Psi_R = \begin{pmatrix} 0 \\ \bar{\eta}^{\dot{\alpha}} \end{pmatrix}. \tag{6.52}$$

The kinetic term for the Weyl spinors involved[7]

$$\mathcal{L}_{\text{kin}} = i\bar{\xi}_{\dot{\beta}}\left(\bar{\sigma}^\mu\right)^{\dot{\beta}\alpha}\partial_\mu\xi_\alpha + i\eta^\alpha\left(\sigma^\mu\right)_{\alpha\dot{\beta}}\partial_\mu\bar{\eta}^{\dot{\beta}} \tag{6.53}$$

[7]Remember that $\bar{\xi}_{\dot{\beta}} = (\xi_\beta)^*$ and $\eta^\alpha = (\bar{\eta}^{\dot{\alpha}})^*$

in terms of the Dirac spinor can be rewritten as

$$\mathcal{L}_{\rm kin} = i\,\bar{\Psi}\,\gamma^\mu\partial_\mu\,\Psi\,, \tag{6.54}$$

where

$$\bar{\Psi} = \Psi^\dagger\gamma^0\,, \tag{6.55}$$

and

$$\gamma^\mu = \begin{pmatrix} 0 & \sigma^\mu \\ \bar{\sigma}^\mu & 0 \end{pmatrix} \tag{6.56}$$

are the Dirac matrices [8] (in the spinor, or Weyl, representation). It is obvious that they satisfy the basic anticommutation relation (the Clifford algebra)

$$\gamma^\mu\gamma^\nu + \gamma^\nu\gamma^\mu = 2g^{\mu\nu} \tag{6.57}$$

and, in addition,

$$(\gamma^0)^\dagger = \gamma^0\,, \qquad (\gamma^i)^\dagger = -\gamma^i\,, \quad i = 1, 2, 3\,. \tag{6.58}$$

If the Dirac spinor is defined as in Eq. (6.51) the charge conjugated spinor Ψ^C can be defined as

$$\Psi^C = \begin{pmatrix} \eta_\alpha \\ \bar{\xi}^{\dot{\alpha}} \end{pmatrix}. \tag{6.59}$$

Imposing the condition that the Dirac bispinor is equal to its charge-conjugated version we get the Majorana bispinor which, then, obviously has the form

$$\lambda = \begin{pmatrix} \eta_\alpha \\ \bar{\eta}^{\dot{\alpha}} \end{pmatrix}. \tag{6.60}$$

We see that the Majorana bispinor describes two degrees of freedom and is equivalent to the Weyl two-component spinor. Thus, given a Weyl spinor we can always construct a Majorana spinor and vice versa. If the Weyl spinor corresponds to the right-handed particle and left-handed antiparticle, the Majorana bispinor describes a "neutral" particle of both polarizations, coinciding with its antiparticle. The Weyl formalism is suitable for the Standard Model. Some models for neutrinos are best written in the Majorana formalism. Both formalisms, Weyl and

[8] Our definition differs in some signs from that in the popular textbook [2]. See page 258.

Majorana, are used in supersymmetric field theory. In the Majorana notation the bilinears one deals with most commonly take the form

$$\bar{\lambda}\lambda = \left(\bar{\eta}_{\dot{\beta}}\bar{\eta}^{\dot{\beta}} + \eta^{\beta}\eta_{\beta}\right), \qquad \bar{\lambda}\gamma_5\lambda = \left(\bar{\eta}_{\dot{\beta}}\bar{\eta}^{\dot{\beta}} - \eta^{\beta}\eta_{\beta}\right), \tag{6.61}$$

and

$$\bar{\lambda}\gamma^{\mu}\lambda = 0\,, \qquad \frac{1}{2}\bar{\lambda}\gamma^{\mu}\gamma^{5}\lambda = \eta^{\alpha}(\sigma^{\mu})_{\alpha\dot{\beta}}\bar{\eta}^{\dot{\beta}}\,. \tag{6.62}$$

Sometimes, instead of the spinor representation, the so-called Majorana representation of gamma matrices is more convenient. In this representation

$$\lambda = \frac{1}{2}\begin{pmatrix} \{\eta + \bar{\eta}\} \\ -i\{\eta - \bar{\eta}\} \end{pmatrix}, \tag{6.63}$$

all gamma matrices are purely imaginary, and the operation of charge conjugation reduces to complex conjugation. In the Majorana representation one can say that the Majorana bispinor is real.

6.5.1 *Two and three dimensions*

In two dimensions, there are no spatial rotations and only one Lorentz boost. The Dirac spinor is a two-component complex spinor,[9]

$$(\Psi_{\rm D})_{\alpha} = \begin{pmatrix} \Psi_1 \\ \Psi_2 \end{pmatrix}, \qquad \Psi^{*}_{1,2} \neq \Psi_{1,2}\,. \tag{6.64}$$

It is convenient to choose two-by-two γ-matrices as

$$\text{2D}: \qquad \gamma^0 = \sigma_2\,, \qquad \gamma^1 = -i\sigma_1\,. \tag{6.65}$$

Then, obviously,

$$\gamma^5 = \gamma^0\gamma^1 = -\sigma_3\,, \quad \gamma^{\mu}\gamma^{\nu} = g^{\mu\nu} + \varepsilon^{\mu\nu}\gamma^5\,. \tag{6.66}$$

Obviously, the chiral spinors in two dimensions have one complex component. Since both γ-matrices in Eq. (6.65) are purely imaginary, the Majorana spinors exist too. They have two real components,

$$(\chi_{\rm M})_{\alpha} = \begin{pmatrix} \chi_1 \\ \chi_2 \end{pmatrix}, \qquad \chi^{*}_{1,2} = \chi_{1,2}\,. \tag{6.67}$$

[9]With the conventions presented below, Ψ_1 is a left-mover while Ψ_2 is a right-mover.

In three dimensions chirality does not exist, as there is no analog of the γ_5 matrix.[10] Three γ-matrices with the Clifford algebra can be chosen as

$$\text{3D}: \qquad \gamma^0 = -\sigma^2\,, \quad \gamma^1 = i\sigma^1\,, \quad \gamma^2 = -i\sigma^3\,, \tag{6.68}$$

where $\sigma^{1,2,3}$ are the Pauli matrices. Thus, in three dimensions the Dirac spinor has two complex components, much in the same way as in two dimensions. Since all three γ-matrices in Eq. (6.68) are purely imaginary, one can define a Majorana spinor. Again, as in two dimensions, it has two real components.

The gamma matrices above satisfy the Dirac condition

$$\{\gamma^\mu \gamma^\nu\} = 2g^{\mu\nu}\,, \qquad \mu, \nu = 0, 1, 2\,. \tag{6.69}$$

With this choice of the gamma matrices, the sigma matrices take the form

$$\sigma^{\mu\nu} = \frac{1}{2}\left[\gamma^\mu \gamma^\nu\right] = \begin{cases} -\sigma^3 \\ -\sigma^1 \\ -i\sigma^2 \end{cases} \tag{6.70}$$

for μ, ν =(0,1), (0,2) and (1,2), respectively. They all are purely real.

In three dimensions there are three Lorentz generators: two boosts and one rotation in the (1,2) plane.

The Majorana spinor is composed of two real components,

$$\psi = \begin{pmatrix} \psi_1 \\ \psi_2 \end{pmatrix}. \tag{6.71}$$

If $\psi_{1,2}$ are defined to be real in one reference frame, for self-consistency they must remain real in any frame. The general transformation from a given frame to another is

$$\psi' = U\psi \equiv \exp\left(-\frac{1}{4}\omega_{\mu\nu}\sigma^{\mu\nu}\right)\psi\,. \tag{6.72}$$

Let us first consider rotation. For rotation by the angle θ we have $\omega^{12} = \omega_{12} = \theta$. Then

$$\psi'_{\rm rot} = \exp\left(\frac{i}{2}\theta\sigma^2\right)\psi = \left(\cos\frac{\theta}{2} + i\sigma^2 \sin\frac{\theta}{2}\right)\psi\,. \tag{6.73}$$

The transformation of rotation in the (1,2) plane is unitary, as it should. Simultaneously it is purely real.

[10] The product $\gamma^0\gamma^1\gamma^2$ reduces to unity. The same statement is valid for any odd number of dimensions.

Now, let us analyze boosts. For boosts we should choose

$$\omega^{0i} = -\omega_{0i} = \phi n^i, \quad \tanh\phi = |\vec{v}| \equiv v, \quad \vec{n} = \vec{v}/v. \tag{6.74}$$

Therefore,

$$\psi'_{\text{boost}} = \exp\left[-\frac{\phi}{2}\left(n^1\sigma^3 + n^2\sigma^1\right)\right]\psi$$

$$= \left(\cosh\frac{\phi}{2} - \left(n^1\sigma^3 + n^2\sigma^1\right)\sinh\frac{\phi}{2}\right)\psi. \tag{6.75}$$

The boost transformation matrix is Hermitean rather than unitary, as it should be. Simultaneously, it is also purely real.

Using the equations above it is not difficult to check, that e.g. $\bar{\psi}\psi = \bar{\psi}'\psi'$ where $\bar{\psi} = \psi^T\gamma^0$.

Appendix 6.1. On Ettore Majorana

Fig. 6.1 Ettore Majorana (1906 – probably 1938) was an Italian theoretical physicist who worked on neutrino masses. He disappeared suddenly under mysterious circumstances while sailing from Naples to Palermo. The Majorana equation and Majorana fermions are named after him.

Appendix 6.2: Derivation of (6.34) from (6.31)

We start from Eq. (6.30),

$$\tilde{\xi} = \exp\left(i\frac{\theta}{2}\vec{n}\vec{\sigma}\right)\xi = \left(\cos\frac{\theta}{2} + i\vec{n}\vec{\sigma}\sin\frac{\theta}{2}\right)\xi\,.$$

Next, we complex-conjugate both sides

$$\tilde{\xi}^* = \left(\cos\frac{\theta}{2} - i\vec{n}\vec{\sigma}^*\sin\frac{\theta}{2}\right)\xi^*$$

and raise indices of $\tilde{\xi}^*$ and ξ^* by applying the matrix $i\sigma_2$ from the left. In this way, we arrive at

$$(i\sigma_2)\tilde{\xi}^* = (i\sigma_2)\left(\cos\frac{\theta}{2} - i\vec{n}\vec{\sigma}^*\sin\frac{\theta}{2}\right)(-i\sigma_2)(i\sigma_2)\xi^*\,.$$

Now we use the fact that $(i\sigma_2)\vec{n}\vec{\sigma}^*(-i\sigma_2) = -\vec{n}\vec{\sigma}$. This completes our derivation.

Appendix 6.3: On Hermann Grassmann and Felix Berezin

Grassmann numbers are named after Hermann Grassmann (they are sometimes called an anticommuting number or anticommuting c-number). Hermann Günther Grassmann (1809-1877) was a German mathematician, linguist, physicist, and publisher. His mathematical work went unnoticed until he was in his sixties.

Beginning in 1827, he studied theology at the University of Berlin, also taking classes in classical languages, philosophy, and literature. He does not appear to have taken courses in mathematics or physics, although mathematics was the field that most interested him. In 1830, he returned to his home town of Stettin (currently in Poland) where he passed the examinations needed to teach mathematics in a gymnasium. His exam results were good enough to allow him to teach only at the lower levels. Around this time, he made his first significant mathematical discoveries, the ones that later led him to the ideas of the "Grassmann numbers" (1844). In his fundamental work *Die lineale Ausdehnungslehre, ein neuer Zweig der Mathematik* published in 1844 he introduced the concepts of linear spaces, Grassmann algebra and manifolds that are currently known as Grassmannians.

In 1834, Grassmann began teaching mathematics at the Gewerbeschule in Berlin. A year later, he returned to Stettin to teach

Fig. 6.2 Hermann Günther Grassmann.

mathematics, physics, German, Latin, and religious studies at a new school, the Otto Schule. Over the next four years, Grassmann passed examinations enabling him to teach mathematics, physics, chemistry, and mineralogy at all secondary school levels. In 1847, he was made an "Oberlehrer" or head teacher. In 1852, he was appointed to his late father's position at the Stettin Gymnasium, thereby acquiring the title of Professor. He never made it to the university level. Grassmann's most important mathematical work was done during his spare time.

The Grassmann numbers was a revolutionary concept, far ahead of its time to be appreciated.

*****

Felix Berezin (1931-1980)was an outstanding Soviet mathematician who in the 1960s and 70s was the driving force behind the emergence of the branch of mathematics now known as supermathematics. The integral over the anticommuting Grassmann variables he introduced in the 1960s laid the foundation for the path integral formulation of quantum field theory with fermions, the heart of modern supersymmetric field theories and superstrings. The Berezin integral is named for him, as is the closely related construction of the Berezinian which may be regarded as the superanalog of the determinant.

Fig. 6.3 Felix Berezin.

Berezin studied at Moscow State University, but was not admitted to the graduate school there on account of his Jewish origin. For the next three years, Berezin taught mathematics at a Moscow high school. He continued to study mathematical physics in his spare time under the guidance of Israel Gelfand. After Khrushchev's liberalization, he managed to get access to the Department of Mathematics at the Moscow State University. Felix Berezin died in an accident: he drowned during a hiking expedition in Soviet Far East. (For more details see [3].)

References

[1] Felix Berezin, *The Method of Second Quantization*, (Academic Press, 1966). For the last Edition see Elsevier, 2012.

[2] E. M. Lifshitz, V. B. Berestetski, L. P. Pitaevskii, *Quantum Electrodynamics* (Landau Course of Theoretical Physics, Vol. 4), Second Edition (Pergamon, 1982).

[3] *Felix Berezin: Life and Death of the Mastermind of Supermathematics*, Ed. M. Shifman, (World Scientific, Singapore, 2007).

Chapter 7

Lecture 7. Fermion Fields in Path Integrals

Fermion determinants. Fermions at finite temperature (a simple example).

7.1 Dirac fields in path integrals

The general strategy is the same as for the bosonic fields except that all expansion coefficients are now represented by Grassmann numbers. In other words, the fermion field is a function of space-time defined on Grassmann numbers. The latter carry appropriate spinor indices. For instance, in four dimensions the Dirac field will carry a spinorial index taking four distinct values, the Weyl field a two-valued index, while in two dimensions the Dirac field will carry an index taking two distinct values, and the Weyl field no indices.

Let us take an arbitrary set of orthonormal basis c-numeric functions

$$\{\phi_i(x)\} \to \int dx\, \phi_i^\dagger(x)\phi_j(x) = \delta_{ij}\,. \tag{7.1}$$

The choice of the basis is dictated by convenience. Then [1]

$$\psi_\alpha(x) = \sum_i \psi_{i\,\alpha}\phi_i(x)\,, \tag{7.2}$$

where α is the spinorial index and the coefficients $\psi_{i\,\alpha}$ are Grassmann numbers. The index i is the label for the number of the mode. Let us choose an arbitrary operator (an infinite-dimensional matrix) O which can contain a differential operator, for instance $O = i\partial_\mu\,(\gamma^\mu))_{\alpha\beta}$.

[1] For simplicity, I assume that one and the same basis $\{\phi^i(x)\}$ is used for all values of the spinorial index α. This requirement can be generalized.

Using Eq. (7.2) one can write

$$\int dx \bar{\psi} O \psi = \sum_{i,\alpha,\beta} \bar{\psi}_{i\,\alpha} \left(\int dx\, \phi_i^\dagger O_{\alpha\beta} \phi_j \right) \psi_{j\,\beta}$$

$$= \sum_{i,\alpha,\beta} \bar{\psi}_{i\,\alpha} O_{i\alpha\, j\beta}\, \psi_{j\,\beta} \tag{7.3}$$

where

$$O_{i\alpha\, j\beta} \equiv \int dx\, \phi_{i\,\alpha}^\dagger O_{\alpha\beta} \phi_{j\,\beta}\,. \tag{7.4}$$

Therefore, using the results of Chapter 6 we conclude that the path integral

$$\int \mathcal{D}\bar{\psi} \mathcal{D}\psi \exp\left(- \int dx \bar{\psi} O \psi \right) = \det O_{i\,\alpha;\, j\,\beta}\,. \tag{7.5}$$

For calculating the fermion partition function (state sum) Z we choose (in the Minkowski space)

$$O = -i\,(i\partial_\mu \gamma^\mu - m) \tag{7.6}$$

(see Sec. 2.3) and arrive at

$$Z = \det\,(-i)\,(i\partial_\mu \gamma^\mu - m)\,. \tag{7.7}$$

The factor $-i$ in front of the expression in the brackets in (7.7) can be omitted, since it is a number that can be included in normalization.

Equations (7.6) and (7.7) refer to free fermions. In QED and Yang-Mills theory fermions interact through their coupling to photons or non-Abelian gauge fields. Hence, Eqs. (7.6) and (7.7) are replaced by

$$O = -i\,(iD_\mu \gamma^\mu - m)\,, \qquad Z = \det\,(-i)\,(iD_\mu \gamma^\mu - m)\,, \tag{7.8}$$

i.e. the partial derivatives are replaced by the covariant derivatives.

Similarly to the bosonic case, we can define the fermion propagator in the Minkowski space as

$$\langle 0 \left| T\{\psi(x_1)\bar{\psi}(x_2)\} \right| 0 \rangle$$

$$= \frac{\int \mathcal{D}\bar{\psi}\mathcal{D}\psi \exp{(iS)}\; \psi(x_1)\bar{\psi}(x_2)}{\int \mathcal{D}\bar{\psi}\mathcal{D}\psi\; \exp{(iS)}}\,,$$

$$S = \int dx\; (\bar{\psi}\, i\partial_\mu \gamma^\mu \psi - m\bar{\psi}\psi) \equiv \int dx\; (\bar{\psi} D_\psi \psi)\,. \tag{7.9}$$

Here the operator D_ψ is defined as

$$D_\psi = i\partial_\mu \gamma^\mu - m\,. \tag{7.10}$$

In order to calculate the fermion two-point function above let us add to the action the source terms,

$$\delta S_\psi = \int dx\, (\bar{\eta}\psi + \bar{\psi}\eta)\,, \tag{7.11}$$

where $\bar{\eta}(x)$ and $\eta(x)$ are Grassmann source functions. The shift of the variables in the path integral to be performed in order to eliminate the linear in ψ terms is

$$\psi(x) = \psi'(x) - D_\psi^{-1}\eta\,, \qquad \bar{\psi}(x) = \bar{\psi}'(x) - \bar{\eta}D_\psi^{-1}\,. \tag{7.12}$$

Then the Z factor takes the form

$$\begin{aligned} Z[\eta,\bar{\eta}] &= \int \mathcal{D}\bar{\psi}\mathcal{D}\psi \, \exp\left[i\,(S + \bar{\eta}\psi + \bar{\psi}\eta)\right] \\ &= Z[\eta = \bar{\eta} = 0] \exp\left[-i\int dx \int dy\, (\bar{\eta}(x) D_\psi^{-1}\eta(y))\right]. \end{aligned} \tag{7.13}$$

To get the two-point function from (7.13) we apply the functional derivative twice,

$$\begin{aligned} \langle 0\,|T\{\psi(x_1)\bar{\psi}(x_2)\}|\,0\rangle &= Z[\eta = \bar{\eta} = 0]^{-1} \\ &\times \left[i\frac{\delta}{\delta\bar{\eta}(x_1)}\right]\left[i\frac{\delta}{\delta\eta(x_2)}\right] Z \\ &= iD_\psi^{-1}(x_1, x_2)\,. \end{aligned} \tag{7.14}$$

In the momentum space, we get the standard propagator

$$\frac{i}{p_\mu\gamma^\mu - m}\,, \tag{7.15}$$

the same as we could obtain using the canonic quantization method. Note that the ordering of all η factors, ψ fields and derivatives over η is important because of the anticommuting nature of the Grassmann numbers.

As was mentioned in the title in the very beginning of this Chapter, all equations above refer to the Dirac fermions.

***

Home assignment before beginning Sec. 7.2: please, re-read Sec. 5.2, page 42, which treats the bosonic theory at finite temperature.

7.2 Fermions at finite temperature

Now let us see how this general theory works in applications. We will consider a (simplified) partition function in fermion quantum mechanics (generalization to field theory is straightforward), described by the Lagrangian

$$\mathcal{L} = i\bar{\psi}\dot{\psi} - m\bar{\psi}\psi\,, \tag{7.16}$$

where the "field" ψ (better to say, the Grassmann variable ψ) and $\bar{\psi}$ depend only on t, i.e. the spacial coordinates are reduced.[2] This is sufficient for our purposes and, at the same time, is sufficiently generic.

As on page 42, we will pass to Euclidean time,

$$t \to -i\tau\,, \qquad \frac{d}{dt} \to i\frac{d}{d\tau}\,. \tag{7.17}$$

In the Euclidean time the Grassmann variables ψ and $\bar{\psi}$ are to be treated as independent.[3] For convenience we can also rescale $\bar{\psi}$ (but not ψ)

$$\bar{\psi} \to -i\bar{\psi}\,, \tag{7.18}$$

although this is certainly not necessary.

Then the Euclidean action takes the form

$$S_E = \int_i^f d\tau\left(-i\bar{\psi}\dot{\psi} - im\bar{\psi}\psi\right), \tag{7.19}$$

while the Euclidean measure in the functional integral is proportional to $\exp\left(-S_E\right)$.

The partition function in quantum mechanics with the discrete spectrum[4] has a very simple definition

$$Z = \sum_n \exp\left(-\beta\langle n|H|n\rangle\right), \qquad \beta = \frac{1}{T}\,, \tag{7.20}$$

where H is the Hamiltonian of the system, $|n\rangle$ is an eigenstate and the sum runs over all eigenstates in the spectrum; β is the inverse temperature. It is convenient to normalize Z assuming that the ground state energy vanishes. This is always possible by subtracting the ground state energy from the Hamiltonian.

[2]The so-called "one-dimensional field theory."

[3]Mathematicians explain that this is due to the absence of the so-called *involution* operation in the Euclidean space. For physicists it is enough to say that $\bar{\psi}$ and ψ cannot be related by the Hermitean conjugation.

[4]That's why I mentioned above that we will be dealing with a simplified example.

In Sec. 5.2 we saw that the calculation of the partition function in the bosonic case requires periodic boundary conditions, cf. Eq. (5.24). In other words, we can say that instead of working in the four-dimensional space R^4 we pass to the cylindric geometry $R^3 \times S^1$. I will show below that in the fermion case to describe thermal physics the boundary conditions should be *anti-periodic*. If we use the periodic boundary condition for fermions we will get not Eq. (7.20), but something else.[5] This is due to the anticommuting nature of the integration variables in the path integral.

All systems with a finite number of Grassmann variables can be implemented in terms of matrix quantum mechanics. Indeed, the Lagrangian (7.16) implies

$$H = \psi \frac{\delta \mathcal{L}}{\delta \dot{\psi}} - \mathcal{L} = m\bar{\psi}\psi\,, \qquad \{\psi\ \bar{\psi}\} = \{\bar{\psi}\ \psi\} = 1\,, \tag{7.21}$$

where the anti-commutation relation in (7.21) is the operator relation following from the fact that

$$\frac{\delta \mathcal{L}}{\delta \dot{\psi}} = \pi_\psi\,.$$

Other anticommutators vanish. The above anticommutation relations are realized by the 2×2 matrices $\sigma_\pm$,

$$\sigma_\pm = \frac{\sigma_1 \pm i\sigma_2}{2}\,, \tag{7.22}$$

where $\sigma_{1,2}$ are the Pauli matrices. Indeed, $\{\sigma_+\ \sigma_-\} = 1$. The correspondence is obviously as follows: $\bar{\psi} \leftrightarrow \sigma_+$ and $\psi \leftrightarrow \sigma_-$.

In terms of the above matrices the Hamiltonian (7.21) takes the form

$$H \to \frac{m}{2}\Big([\sigma_+\sigma_-] + \{\sigma_+\sigma_-\}\Big) = \frac{m}{2}(1+\sigma_3)\,. \tag{7.23}$$

The ground state is obviously at zero, so everything is alright with the normalization of the partition function. The only excited state is at $E = m$. The partition function of this system consists of two terms,

$$Z = 1 + e^{-m/T}\,. \tag{7.24}$$

Now, let us see what we get for the same quantity using the path integral in the Euclidean time,

$$Z = \int D\bar{\psi}D\psi \exp\left(-\int d\tau \bar{\psi}\left(-i\frac{d}{d\tau} - im\right)\psi\right) = \det\left(i\frac{d}{d\tau} + im\right). \tag{7.25}$$

[5]In supersymmetric theories this "something else" is referred to as the *Witten index*.

As was mentioned, the anti-symmetric boundary condition $\psi(0) = -\psi(\beta)$ is required at the temperature $T = 1/\beta$. In other words, the basis functions must be chosen under the condition

$$\phi_k(\tau) = \exp\left[i\,(2k+1)\,\pi\tau/\beta\right], \qquad k = 0 \pm 1, \pm 2, ... \tag{7.26}$$

with the following eigenvalues of the operator $\left(i\frac{d}{d\tau} + im\right)$:

$$E_k = -(2k+1)\,\pi/\beta + im\,. \tag{7.27}$$

Taking the eigenvalues pairwise we arrive at

$$Z = \det\left(i\frac{d}{d\tau} + im\right) = \prod_{k=0}^{\infty}(-1)\left[\frac{4\pi^2}{\beta^2}\left(k+\frac{1}{2}\right)^2 + m^2\right]. \tag{7.28}$$

As any infinite product, it should be treated with care. Taken at its face value the product (7.28) diverges and is therefore ill-defined. The necessary condition for its convergence is that at $k \to \infty$ the k-th factor in the product tends to unity fast enough.

To deal with this circumstance we should note that at $T \to 0$ (or $\beta \to \infty$) contribution of all excitations in the partition function must vanish,[6] and the partition function must be saturated by the ground state, which in our convention means that $Z \to 1$ in this limit. We can use this condition to normalize (7.28). Indeed, let us split the product

$$\prod_{k=0}^{\infty}(-1)\left[\frac{4\pi^2}{\beta^2}\left(k+\frac{1}{2}\right)^2 + m^2\right]$$
$$\to \prod_{k=0}^{\infty}(-1)\left[\frac{4\pi^2}{\beta^2}\left(k+\frac{1}{2}\right)^2\right] \times \prod_{k=0}^{\infty}\left[1 + \frac{m^2\beta^2}{4\pi^2\left(k+\frac{1}{2}\right)^2}\right]. \tag{7.29}$$

The first factor on the right-hand side is infinite but m independent, and we can drop it. The second factor reduces to

$$\prod_{k=0}^{\infty}\left[1 + \frac{m^2\beta^2}{4\pi^2\left(k+\frac{1}{2}\right)^2}\right] = \cosh\frac{m\beta}{2}\,. \tag{7.30}$$

The function on the right-hand side is a sum of two exponents. One of them grows at large β. Surprise, surprise... What's the matter? The expression for Z in (7.30) corresponds to a *negative* energy of the

[6] Alternatively one can fix T and consider $m \to \infty$.

ground state. Normalizing to the limit $Z \to 1$ at $\beta \to \infty$ we arrive at

$$Z = 1 + e^{-m\beta}\,, \tag{7.31}$$

to be compared with Eq. (7.24).

What changes if, instead of the anti-periodic boundary condition we chose periodic, $\psi(0) = \psi(\beta)$?

Then the set (7.26) becomes

$$\phi_k(\tau) = \exp\left[i\,(2\pi)\,k\tau/\beta\right], \qquad k = 0 \pm 1, \pm 2, ... \tag{7.32}$$

with the following eigenvalues of the operator $\left(i\frac{d}{d\tau} + im\right)$:

$$E_k = -2k\,\pi/\beta + im\,. \tag{7.33}$$

For the product corresponding to the second factor in (7.29) we obtain

$$\prod_{k=1}^{\infty}\left[1 + \frac{m^2\beta^2}{4\pi^2\,k^2}\right] \tag{7.34}$$

where the $k = 0$ mode is singled out. In contradistinction with (7.30) we obtain now

$$\sinh\frac{m\beta}{2} \to 1 - e^{-m\beta}\,. \tag{7.35}$$

The sign of the second term (i.e. the fermion excitation) is flipped from plus to minus. This is the general feature of non-thermal treatment [7] of fermions in path integrals ("non-thermal treatment" means the periodic boundary condition in the case of fermions). This negative term appears in the so-called Witten index [1].

***

Out of curiosity, let us return to Sec. 5.2 and calculate the temperature partition function for a free complex scalar field (in zero spacial dimensions), with the Lagrangian

$$\mathcal{L} = \dot\phi^\dagger\dot\phi - m^2\phi^\dagger\phi\,. \tag{7.36}$$

The system is obviously reducible to two independent harmonic oscillators, which is seen through the substitution of the complex variable ϕ by two real,

$$\phi(t) = \frac{1}{\sqrt{2}}\left(\eta(t) + i\chi(t)\right), \tag{7.37}$$

[7] Such treatment with the periodic boundary conditions in the fermion case is relevant not for the partition function, but, instead, for the Witten index in supersymmetric theories.

leading us to the Lagrangian

$$\mathcal{L} = \frac{1}{2}\dot{\eta}\dot{\eta} - \frac{1}{2}m^2\eta\eta + \frac{1}{2}\dot{\chi}\dot{\chi} - \frac{1}{2}m^2\chi\chi\,. \tag{7.38}$$

In Euclidean space the action takes the form

$$S_E = \int d\tau \left(\frac{1}{2}(\dot{\eta}\dot{\eta} + m^2\eta\eta + \dot{\chi}\dot{\chi} + m^2\chi\chi)\right). \tag{7.39}$$

Now, the functional integration of $\exp(-S_E)$ over $\mathcal{D}\phi$ and $\mathcal{D}\phi^\dagger$ yields

$$\frac{1}{\det\left(-\frac{d^2}{d\tau^2} + m^2\right)} = \prod_{k=-\infty}^{+\infty} \frac{1}{\left(\frac{2\pi k}{\beta}\right)^2 + m^2}\,, \tag{7.40}$$

where on the right-hand side the periodic boundary condition (i.e. the basis (7.32)) is taken into account. It is convenient to isolate the $k=0$ mode[8] and rewrite the right-hand side of (7.40) as follows:

$$m^2 \prod_{k=1}^{+\infty} \frac{1}{\left[\left(\frac{2\pi k}{\beta}\right)^2 + m^2\right]^2} \to \prod_{k=1}^{+\infty} \frac{1}{\left[1 + \frac{\beta^2 m^2}{4\pi^2 k^2}\right]^2}$$

$$\to \left(\sinh\frac{m\beta}{2}\right)^{-2} \to \left[1 - \exp\left(-m\beta\right)\right]^{-2}$$

$$= 1 + \sum_{k=1}^{\infty}(k+1)\exp\left[-km\beta\right]. \tag{7.41}$$

The ground state is normalized to unity, the first excitation contributes $2\exp\left(-m\beta\right)$, the second excitation $3\exp\left(-2m\beta\right)$ and so on.

If the exponents in (7.41) have a clear-cut interpretation (in the harmonic oscillator each excitation adds m in energy), the occurrence of pre-exponents requires an explanation. They indicate the multiplicity of the given level.

Let us return to Eqs. (7.38) and (7.39) telling us that in fact we deal here with two independent harmonic oscillators. The energy is counted from the ground state, which by definition is placed at zero. Then we have a single ground state (0,0). The numbers in the parentheses indicate in which state the first and the second oscillators are. Let us call (0,0) level zero. At level one we have two options: (0,1) or (1,0).

[8]The reason why this "zero mode" must be isolated and discarded in Eq. (7.40) is discussed in detail in Chapter 22, page 235.

Hence, the factor 2 in front of $\exp(-m\beta)$. This is the first excited state of the system. At level two we have three options (0,2) or (2,0) and (1,1). Hence, the factor 3 in front of $\exp(-2m\beta)$, and so on.

Reference

[1] E. Witten, Nucl. Phys. B **202**, 253 (1982) [Reprinted in *Supersymmetry*, Ed. S. Ferrara (North Holland/World Scientific, Amsterdam-Singapore, 1987), Vol. 1, page 490].

Chapter 8

Lecture 8. Gauge Fields in Path Integral: The Simplest Example of QED

Gauge orbits. The need for gauge fixing. Gauge fixing conditions. Faddeev-Popov determinant. Why we can drop it in QED? The R_ξ gauge in QED.

8.1 Photon fields in path integral

The photon part of the QED Lagrangian can be written as

$$\mathcal{L}_\gamma = -\frac{1}{4} F_{\mu\nu} F^{\mu\nu} . \tag{8.1}$$

The action takes the form

$$\begin{aligned} S &= \frac{1}{2} \int dx \, A^\mu \left(\partial^2 g_{\mu\nu} - \partial_\mu \partial_\nu \right) A^\nu \\ &\to \frac{1}{2} \int dk (2\pi)^{-4} \left[\tilde{A}^\mu(k) \left(-k^2 g_{\mu\nu} + k_\mu k_\nu \right) \tilde{A}^\nu(-k) \right] . \end{aligned} \tag{8.2}$$

The operator $D_A^2 = -\partial^2 g_{\mu\nu} + \partial_\mu \partial_\nu$ cannot be inverted, the inverse D_A^{-2} does not exist!

Indeed, in the momentum space D_A^2 is a four by four matrix,

$$k^2 g_{\mu\nu} - k_\mu k_\nu . \tag{8.3}$$

The matrix in (8.3) cannot be inverted. The easiest way to see this is to align $\vec{k}$ along the z-axis. Then it is not difficult to calculate this matrix. It takes the form

$$\begin{pmatrix} -k_z^2 & 0 & 0 & -k_0 k_z \\ 0 & -k_0^2 + k_z^2 & 0 & 0 \\ 0 & 0 & -k_0^2 + k_z^2 & 0 \\ -k_0 k_z & 0 & 0 & -k_0^2 \end{pmatrix} \tag{8.4}$$

and its determinant

$$k_0^2 k_z^2 (k_0^2 - k_z^2)^2 - k_0^2 k_z^2 (k_0^2 - k_z^2)^2$$

vanishes, i.e. $D_A^{-2} = \infty$.

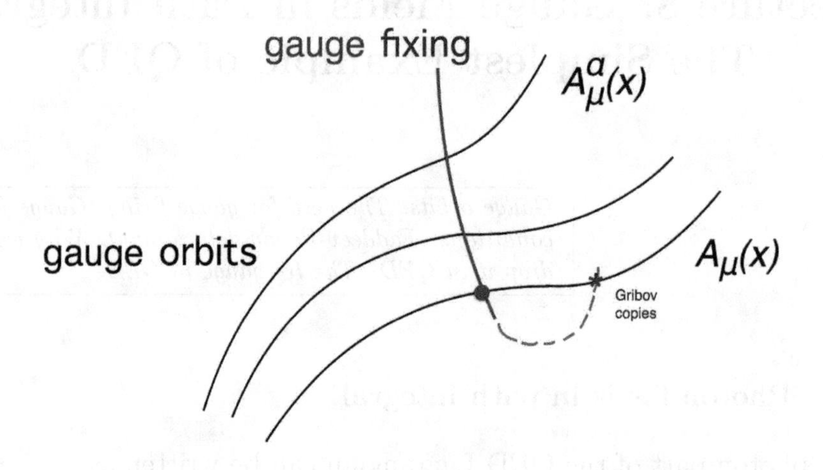

Fig. 8.1 Each field configuration $A_\mu(x)$ is represented by an infinite number of points on the gauge orbit. In the path integral each gauge orbit must be represented once. This is done by *gauge fixing*, i.e. imposing an appropriate gauge condition, e.g. $G(A) = 0$ for all x, as in Eq. (8.8). In non-Abelian theories some conventional gauge-fixing conditions may lead to more than a single intersection of the red line with a given gauge orbit at large values of the A^a_μ fields. This phenomenon is called *Gribov copies.* The Gribov copies do not reveal themselves in perturbation theory, however. To avoid multiple counting one should choose a single "point" from the gauge orbit (see the closed circle).

This "disaster" is due to the phenomenon of gauge orbits, see Fig. 8.1. Because of the gauge redundancy, every physical field configuration is represented in the functional integral infinitely many times (a continuous infinity). Let me rephrase this statement due to its importance. In the non-gauge theories, such as $\lambda\phi^4$, every point in the space of fields is represented by a smooth function $\phi(x)$. This is not the case in gauge theories. Indeed, if we choose a field configuration $A_\mu(x)$ its arbitrary gauge transform is physically one and the same point in the space of fields. Therefore, the space of fields consists of equivalence classes $\{A^\alpha_\mu(x)\}$,

$$A^\alpha_\mu(x) \equiv A_\mu(x) + \partial_\mu \alpha(x) \,, \tag{8.5}$$

rather than of individual fields $A_\mu(x)$. Here the superscript α denotes a set of all possible gauge transformations. The above equivalence class

$\{A_\mu^\alpha(x)\}$ is the same as the gauge orbit. I represented it graphically in Fig. 8.1. Correspondingly, the path integrals, e.g.

$$Z = \int \mathcal{D}A_\mu(x) \exp\left(iS_{\text{gauge}}\right) \tag{8.6}$$

should be understood as

$$Z = \int \mathcal{D}\{A_\mu^\alpha(x)\} \exp\left(iS_{\text{gauge}}\right). \tag{8.7}$$

Moving along the gauge orbit, we keep the integrand constant because the action of gauge theories S_{gauge} is gauge independent. As a result, the integral (8.6) is proportional to an infinite "gauge volume" $\infty^{V(G)}$, where infinity represents the space-time continuum.

What could be done? We can pick up a single representative from each gauge orbit, and integrate in the functional integral over the equivalence classes, counting only chosen representatives. A physicist would say that we should move "perpendicular" to the stream of orbits. A mathematician would say that we should define a coset space of the group orbits. This space is shown in Fig. 8.1 by a red line.

Thus, to eliminate multiple counting we need to introduce a condition that precludes us from moving along the gauge orbit and will force us to move in the "perpendicular" directions (the space of fields is infinitely-dimensional; therefore the above statements are just graphic simplifications). One of the standard gauge conditions which forbids (at least in QED) the motion along the gauge orbit starting from $A_\mu = 0$ is[1]

$$G(A) \equiv \partial_\mu A^\mu(x) = 0. \tag{8.8}$$

Using Eq. (8.5) we obtain

$$\partial^\mu A_\mu^\alpha = \partial^\mu A_\mu + \partial^2 \alpha \tag{8.9}$$

implying

$$\Box \alpha(x) = 0. \tag{8.10}$$

The Laplace equation above defines the so-called harmonic functions. If we impose the boundary condition at infinity

$$\alpha \to 0 \text{ at } |x| \to \infty \tag{8.11}$$

and require no singularities at finite x, then the only solution is $\alpha = const = 0$. This indeed excludes the motion along the gauge orbit.

[1]In perturbation theory only small fields close to $A_\mu = 0$ are important.

Thus, let us try to impose (8.8) at every space-time point. To this end we want to change the measure in the path integral from $\int \mathcal{D}A...$ to

$$\int \mathcal{D}A(x)...\delta(G(A(x)))\,.$$

The delta function above should be understood as a functional delta function, something like

$$\prod_{\forall x} \delta(G(A(x))) = \delta^{\infty}(G(A(x)))\,. \tag{8.12}$$

It must be assumed that the condition

$$G(A_\mu^\alpha) = 0 \tag{8.13}$$

has a unique solution at $\alpha = \alpha_*$. This can be achieved not for every gauge condition, see below.

Insertion of this factor should not change the path integral, however. A device to achieve this goal was invented by Faddeev and Popov (who developed the formalism of the path integral quantization of the gauge fields, see Appendix 8.1 on page 77). The so-called Faddeev-Popov **1** ("unity") is

$$\mathbf{1} = \int \mathcal{D}\alpha(x) \det\left[\frac{\delta G(A^\alpha)}{\delta\alpha(x)}\right]_{\alpha_*} \delta^{\infty}(G(A^\alpha))\,, \tag{8.14}$$

where the determinant above is evaluated at the point $\alpha_*(x)$ such that $G(A^{\alpha_*}(x)) = 0$. Remember, A^α is defined in (8.5). In other words, for each $A(x)$ we find $\alpha^*(x)$ such that $G(A^{\alpha^*(x)}) = 0$. The determinant above is obviously gauge invariant.

Equation (8.14) is a functional generalization of the following finite-dimensional identity for an n-component vector $\vec{g} = \{g_1, g_2, ..., g_n\}$:

$$\left[\prod_{i=1}^{n} \int da_i \delta(g_i(a))\right] \det||\partial g_i/\partial a_k|| = \prod_{i=1}^{n} \int dg_i\, \delta(g_i) = 1\,, \tag{8.15}$$

Each of the n components of the vector $\vec{g}$ is assumed to depend on all a_i where $i = 1, 2, .., n$. It is also assumed that the zero of $\vec{g}$ corresponds to a unique point in $\{a\}$ space. The role of $a_1, a_2, ...$ is played by $\alpha(x_1), \alpha(x_2), ...$ while the role of $g_i(\alpha)$ by $G(A^\alpha)$ at x_1, x_2 and so on.

Next, we take the defining functional integral

$$\int \mathcal{D}A e^{iS[A]} \tag{8.16}$$

and insert in the integral the Faddeev-Popov unity (8.14),

$$\int \mathcal{D}A e^{iS[A]} = \int \mathcal{D}A(x)\mathcal{D}\alpha(x) \det\left[\frac{\delta G(A^\alpha)}{\delta\alpha(x)}\right]_{\alpha*} \delta^\infty(G(A^\alpha(x))\, e^{iS[A]} \,. \tag{8.17}$$

With our choice of the gauge condition

$$G(A^\alpha) = \partial^\mu \left[A_\mu(x) + \partial_\mu \alpha(x)\right]$$

the determinant in (8.14) becomes

$$\det\left[\frac{\delta G(A^\alpha(x))}{\delta\alpha(y)}\right] \propto \det\left[\partial^2 \delta(x-y)\right] . \tag{8.18}$$

What is important for our purposes is the fact that the determinant in (8.18) does not depend on A and therefore can be moved from $\int \mathcal{D}A...$ outside the integral. (Warning: this is not the case for the *non-Abelian gauge theories!*). Moreover, because it is A independent, it will be absorbed by normalization factors.

Now, let us take into account the identities

$$S[A] = S[A^\alpha], \text{ and } \mathcal{D}A(x) = \mathcal{D}A^\alpha(x) , \tag{8.19}$$

substitute them in (8.17), and change the integration variable from $A(x)$ to $A^\alpha(x)$. Then the path integral (8.16) takes the form

$$\int \mathcal{D}A e^{iS[A]} = \int \mathcal{D}\alpha(x) \det\left[\frac{\delta G(A^\alpha)}{\delta\alpha(x)}\right]_{\alpha*} \int \mathcal{D}A^\alpha(x) \delta^\infty(G(A^\alpha(x))\, e^{iS[A^\alpha]} \,. \tag{8.20}$$

Let us rename

$$A^\alpha \to A \,.$$

After doing so, the integral on the right-hand side of (8.20) splits into two factors: one is an overall normalization

$$\int \mathcal{D}\alpha(x) \det\left[\frac{\delta G(A)}{\delta\alpha(x)}\right]_{\alpha=0} \tag{8.21}$$

which can be dropped [2] and the second, well-defined, factor

$$\int \mathcal{D}A e^{iS[A]} \to \int \mathcal{D}A(x)\, \delta^\infty(G(A))\, e^{iS[A]} \tag{8.22}$$

Note that in Yang-Mills theories, the factor we have just discarded cannot be absorbed in normalization and tossed away.

[2] It is infinite, of course. But still, it is an overall normalization which can be dropped because it is A *independent*, see above.

*****

To implement the gauge fixing in a more general and more familiar form, let us introduce a slightly different gauge condition (which does not change anything in the above derivation). Let us assume that (cf. Eq. (8.8))

$$(G(A(x)) = \partial_\mu A^\mu - \omega(x) \tag{8.23}$$

where $\omega(x)$ is an arbitrary scalar function. The determinant

$$\left[\frac{\delta G(A^\alpha(x))}{\delta\alpha(y)}\right]$$

is the same as in Eq. (8.18). Thus, now instead of (8.22) we have

$$\int \mathcal{D}A(x)e^{iS[A]} \to \int \mathcal{D}A(x)e^{iS[A]}\delta^\infty(\partial_\mu A^\mu - \omega(x)). \tag{8.24}$$

The above equation is valid for any ω and so is any combination of integrals on the right-hand side with different ω's. In particular, we can average over all ω's with the Gaussian weight factor

$$N(\xi)\int \mathcal{D}\omega(x)\exp\left(-i\int d^4x\,\frac{\omega^2}{2\xi}\right), \tag{8.25}$$

where ξ is an arbitrary constant parameter and $N(\xi)$ is an irrelevant normalizing factor.

Applying (8.25) to (8.24) we arrive at

$$\int \mathcal{D}A(x)e^{iS[A]} \stackrel{\text{gf}}{\to} \int \mathcal{D}A(x)e^{iS[A]}\exp\left[-i\int d^4x\,\frac{1}{2\xi}\left(\partial_\mu A^\mu\right)^2\right] \tag{8.26}$$

The term in the square brackets is the gauge-fixing extra term in the Lagrangian.

Now, the quadratic form in Eq. (8.2) becomes

$$-\frac{1}{2}\int\frac{d^4k}{(2\pi)^4}\left\{\tilde{A}^\mu(k)\,k^2\left[g_{\mu\nu}-\left(1-\frac{1}{\xi}\right)\frac{k_\mu k_\nu}{k^2}\right]\tilde{A}^\nu(-k)\right\}. \tag{8.27}$$

Inverting the matrix in the square brackets we arrive at the following photon Green's function:

$$D_{\alpha\beta}(k) = \frac{-i}{k^2}\left[g^{\alpha\beta} - (1-\xi)\frac{k_\alpha k_\beta}{k^2}\right]. \tag{8.28}$$

Two gauges are most popular: $\xi = 1$ (the Feynman gauge) and $\xi = 0$ (the Landau gauge, sometimes referred to as the Lorentz gauge).

Fig. 8.2 Ludwig Faddeev.

Appendix 8.1: On L. D. Faddeev

Ludwig Faddeev (1934-2017) is a Russian theoretical physicist and mathematician, famous for the discovery of the Faddeev equations in the theory of the quantum mechanical three-body problem and for the development of path integral methods in the quantization of non-abelian gauge field theories, including the introduction (with Victor Popov) of the Faddeev-Popov ghosts. He led the Leningrad mathematical school, in which he with his students developed the quantum inverse scattering method and the framework for quantization of solitons. He also made a crucial contribution to the study of quantum integrable systems in one space and one time dimension.

L. Faddeev was born in Leningrad into a family of mathematicians. His father, Dmitry Faddeev, was a well known algebraist, while his mother was known for her work in numerical linear algebra. Faddeev graduated from Leningrad University in 1956 and earned his doctorate in 1959.

Chapter 9

Lecture 9. Non-Abelian Gauge Fields in Path Integrals

> *Faddeev-Popov ghosts in covariant gauges in Yang-Mills theory. Their physical meaning. Some properties of the group generators in $SU(N)$.*

I will begin this lecture from rephrasing the derivation for the Abelian case (discussed in Chapter 8) in the hope that this rephrasing can facilitate the understanding of subtle points of Yang-Mills quantization. To this end, I find it helpful to pass to the Fourier-transformed formulation, assuming that the (four-dimensional) momentum space is somehow discretized.

In the Abelian case we can write

$$A_\mu(x) = \sum_k e^{ikx} a_\mu^k \,, \quad F_{\mu\nu}^k = i(k_\mu a_\nu^k - k_\nu a_\mu^k) \tag{9.1}$$

and

$$S = -\frac{1}{4} \sum_k F_{\mu\nu}^k F^{\mu\nu,\,-k} \,. \tag{9.2}$$

Then, naively, the path integral will take the form

$$\int \mathcal{D}A e^{i \int dx L[A(x)]} \to \int \prod_{\forall k} da_\mu^k \exp\left(-\frac{i}{4} \sum_k F_{\mu\nu}^k F^{\mu\nu,\,-k}\right) . \tag{9.3}$$

Equation (9.3) does not take into account the gauge freedom. In the functional integration we want to pass from field configurations to equivalence classes identifying all field configurations inside the class,

$$A^\mu(x) + \partial^\mu \alpha \,.$$

In the Fourier-transformed representation, we identify

$$a_\mu(k)^\alpha = a_\mu(k) + ik_\mu \alpha_k \,, \qquad \alpha(x) = \sum_k e^{ikx} \alpha_k \,. \tag{9.4}$$

For what follows it is convenient to split $a_\mu(k)$ into two parts, transversal and longitudinal,

$$a_\mu(k) \equiv a_\mu(k)_{||} + a_\mu(k)_\perp$$

$$a_\mu(k)_\perp = \left(g_{\mu\nu} - \frac{k_\mu k_\nu}{k^2}\right) a_\nu(k)\,, \quad a_\mu(k)_{||} = \frac{k_\mu k_\nu}{k^2}\, a_\nu(k) \tag{9.5}$$

Moreover,

$$\delta^\infty\left(\partial_\mu(A^\mu(x) + \partial^\mu \alpha)\right) \to \prod_k \delta(ik^\mu a_\mu^k - k^2\alpha^k) \tag{9.6}$$

$$= \prod_k \delta\left(\alpha^k - \frac{ik^\mu a_\mu^k}{k^2}\right)\frac{1}{k^2}\,, \tag{9.7}$$

and α^{k*} saturating the δ function in (9.7) is

$$\alpha^{k*} = \frac{ik^\mu a_\mu^k}{k^2}\,, \quad \text{implying } \left(a_\mu^{\alpha^*}\right)_{||} = 0\,. \tag{9.8}$$

The gauge fixing term is defined in terms of the function

$$G(A^\alpha) = ik^\mu a_\mu^k - k^2\alpha^k \tag{9.9}$$

Then the corresponding determinant is

$$\det\left|\left|\frac{\delta G}{\delta\alpha}\right|\right| = \prod_k k^2\,. \tag{9.10}$$

The Faddeev-Popov identity takes the form

$$\mathbf{1} = \det\left|\left|\frac{\delta G}{\delta\alpha}\right|\right| \int \prod_k d\alpha_k \delta(ik^\mu a_\mu^k - k^2\alpha^k) = \int d\alpha_k \prod_k \delta\left(\alpha^k - \frac{ik^\mu a_\mu^k}{k^2}\right). \tag{9.11}$$

The integral on the right-hand side of (9.11) produces two effects: (i) the overall normalization; (ii) enforcing that $\left(a_\mu^k\right)_{||} = 0$. Thus, we arrive at

$$\int \mathcal{D}A_\mu e^{i\int dx L[A(x)]} \to \int \prod_{\forall k} d\left(a_\mu^k\right)_\perp \exp\left(-\frac{i}{4}\sum_k F_{\mu\nu}^k F^{\mu\nu,-k}\right). \tag{9.12}$$

Equation (9.12) is parallel to Eq. (8.22).

*****

Now, we can proceed to the non-Abelian case. The difference is not large, but crucial. In the Abelian case

$$G[A^\alpha(x)] = \partial_\mu A^{\mu,\alpha}(x) = \partial_\mu A^\mu(x) + \Box\alpha(x)\,. \tag{9.13}$$

Correspondingly,

$$\frac{\delta G[A^\alpha(x)]}{\delta\alpha(y)} = \partial^2\delta(x-y)\,. \tag{9.14}$$

The operator on the right-hand side is A independent, and, therefore, its determinant can be absorbed into the overall normalization.

Now, we will explore the gauge-transformed A^α for the Yang-Mills fields. To this end, we will need Eq. (1.26). I will reproduce it below:

$$A^{\mu\,a}_{\text{gt}}(x)\,T^a = U(x)\left(A^{\mu\,a}(x)\right)T^a\,U^\dagger(x) + iU(x)\left(\partial^\mu U(x)^\dagger\right)\,. \tag{9.15}$$

This is a general gauge transformation. The last (inhomogeneous) term on the right-hand side is an analog of the Abelian gauge transformation. The first (homogeneous) term on the right-hand side of (9.15) was absent in the Abelian case.

Next, we assume that U is close to unity (i.e. expand in the transformation parameter). This will be sufficient for perturbation theory. Then

$$U(x) = 1 + i\alpha^b(x)T^b + \ldots\,, \qquad b = 1,2,\ldots,N^2-1\,. \tag{9.16}$$

The first term in (9.15) takes the form

$$\begin{aligned} U(x)\left(A^{\mu\,a}(x)\right)T^a\,U^\dagger(x) &= A^{\mu\,a}(x)T^a + i\alpha^b(x)A^{\mu\,a}(x)\left[T^bT^a\right] \\ &= A^{\mu\,a}(x)T^a + i\alpha^b(x)A^{\mu\,a}(x)\,if^{bac}\,T^c\,. \end{aligned} \tag{9.17}$$

The second term in (9.15) takes the form

$$iU(x)\left(\partial^\mu U(x)^\dagger\right) = \partial^\mu\alpha^a\,T^a\,. \tag{9.18}$$

Combining (9.17) and (9.18) we arrive at

$$\begin{aligned} \left(A^{\mu\,a}(x)\right)^\alpha &= A^{\mu\,a}(x) + \partial^\mu\alpha^a(x) + f^{abc}A^{\mu\,b}(x)\alpha^c(x) \\ &= A^{\mu\,a}(x) + D^\mu\alpha^a(x)\,, \end{aligned} \tag{9.19}$$

where D^μ in the second line is the covariant derivative.

Just as in the Abelian case, the integral over the gauge orbits can be factored out of the path integral, leaving us with an extra determinant in the integrand

$$\int \mathcal{D}A^{\mu\,a}\, e^{iS[A]} \to \int \mathcal{D}\alpha^a(x) \int \mathcal{D}A^{\mu\,a}\, e^{iS[A]}\, \delta^\infty(G[A^\alpha])$$
$$\times \det \left|\left| \frac{\delta G(A^\alpha)}{\delta \alpha(y)} \right|\right| . \tag{9.20}$$

The factor in the top line in (9.20) is infinite since it produces the overall volume of the gauge orbit. This is in parallel with the same effect as in the Abelian theory. However, let us have a closer look at the determinant in the bottom line,

$$\det \left|\left| \frac{\delta G(A^\alpha)}{\delta \alpha(y)} \right|\right| .$$

The operator

$$\frac{\delta G(A^\alpha(x))}{\delta \alpha(y)} \equiv \frac{\delta\left[\partial_\mu\left(A^{\mu\,a}(x) + D^\mu \alpha^a(x)\right)\right]}{\delta \alpha(y)} = \partial^\mu D_\mu\, \delta(x-y)\,. \tag{9.21}$$

Now, it depends on the fields $A^a_\mu(x)$ through the covariant derivative on the right-hand side. This is a major and very crucial distinction with regards to the Abelian theory. The functional determinant cannot be absorbed into the overall normalization.

Faddeev and Popov invented a simple representation for the above determinant.[1] Namely

$$\det \partial^\mu D_\mu \equiv \int \mathcal{D}c\, \mathcal{D}\bar{c} \exp\left[i \int d^4x \bar{c}(-\partial^\mu D_\mu) c\right] \tag{9.22}$$

where $c, \bar{c}$ are complex *scalar* fields with the Fermi statistic in the *adjoint* representation of the gauge group, i.e.

$$(D_\mu c)^a = \left(\delta^{ac}\partial_\mu + f^{abc} A^b_\mu\right) c^c\,, \tag{9.23}$$

in much the same way as for the gauge bosons themselves. Were they *bona fide* scalar fields, they should have obeyed the Bose statistic, but then, instead of the determinant on the left-hand side, we would get $1/\det \partial^\mu D_\mu$. The fields c, $\bar{c}$ are unphysical; they appear only in loops; their only *raison d'etre* is to cancel unphysical polarizations of the gauge fields.[2] That's why they are called the Faddeev-Popov ghosts.

[1]They followed Berezin's suggestion.
[2]We will explicitly see this later.

The ghost addition to the Lagrangian takes the form

$$\mathcal{L}_{\rm ghost} = \partial_\mu \bar{c}\, D^\mu c\,. \tag{9.24}$$

Please, observe the asymmetry in the derivatives: one is just a regular partial derivative, the other one is the covariant derivative.

Treatments of non-Abelian theories in the path integral formulation can be found in many textbooks. The curious reader is referred to [1; 2; 3; 4] for further details which will not be used in our course, though.

Appendix 9.1: Properties of group generators and the Casimir coefficients

The generators in the representation R obey the standard commutation relations

$$[T^a_R\,,\,T^b_R] = i f^{abc}\, T^c_R\,, \tag{9.25}$$

and are normalized in a conventional manner,

$$T^a_R T^a_R = C_2(R),\;\; {\rm Tr}\,(T^a_R T^b_R) = T(R)\,\delta^{ab},\;\; T(R) = C_2(R)\,\frac{\dim(R)}{\dim({\rm adj})}, \tag{9.26}$$

where $C_2(R)$ is the quadratic Casimir operator and $T(R)$ is called (one half of) the Dynkin index in the mathematical literature. Sometimes $T({\rm adj}) \equiv T_G$ is referred to as the dual Coxeter number. For the fundamental representation $T({\rm fund}) = 1/2$ while $T({\rm adj}) = N$. Note that the generators of a given complex representation R are related to those of the complex conjugated representation $\bar{R}$ by the formula

$$\bar{T}^a = -\tilde{T}^a = -T^{a\,*}\,, \tag{9.27}$$

where the tilde denotes the transposed matrix. The Casimir coefficients for some representations of SU(N) are compiled below.

Representations→ / Casimirs↓	Fundamental	Adjoint	2-index A	2-index S
$T(R)$	$\frac{1}{2}$	N	$\frac{N-2}{2}$	$\frac{N+2}{2}$
$C_2(R)$	$\frac{N^2-1}{2N}$	N	$\frac{(N-2)(N+1)}{N}$	$\frac{(N+2)(N-1)}{N}$

References

[1] L. D. Faddeev and A. A. Slavnov, *Gauge Fields: An Introduction to Quantum Theory*, Second Edition, (CRC Press, 1993).
[2] Pierre Ramond, *Field Theory: A Modern Primer*, Second Edition, (Addison-Wesley, 1990).
[3] Stefan Pokorski, *Gauge Field Theories*, (Cambridge University Press, 1987).
[4] M. Peskin and D. Schroeder, *An Introduction to Quantum Field Theory*, (Westview Press, 1995).

Chapter 10

How to Calculate Charge Renormalization

> *The Wilsonean paradigm. Various contributions to the gauge coupling renormalization: spin-zero (bosons) and spin-1/2 (fermions). A (relatively) simple way. Pauli–Villars regularization. Useful expressions for massless propagators in the coordinate space.*

Today we will begin a new topic. My final goal is to teach you how to calculate the coupling constant renormalization (a.k.a running coupling) in Yang-Mills theory. It will take us some time; we will start from a simple set-up (quantum electrodynamics, QED) and gradually move toward more complicated things.

If you read relatively old texts you will see that a lot of attention is paid to the so-called "mystery" of renormalization, "infinities" and what to do with them. Renormalizability vs. non-renormalizability was a major headache for theorists for 30 or 40 years. Now the paradigm has totally changed. It is understood that none of field theories is fundamental "till the very end." In other words, at short distances (large momenta) field theory emerges as a low-energy limit of some "more fundamental theory" presumably (I emphasize, *presumably*) including quantum gravity. Perhaps, this "more fundamental theory" will be string theory. Well... who knows, the future's not ours to see...

Currently the starting point is as follows. The original action of field theory is assumed to be formulated at a certain point (at a large but finite Euclidean momentum which is equivalent to a short but finite distance) and includes all *relevant operators* with constants that are referred to as "bare" constants. Then one starts evolving it to smaller momenta (larger distances), see Fig. 10.1. This evolution is called "Renormalization Group flow" (also, RG flow, or Wilsonian RG approach). We will have a special lecture on it. In this evolution, covering the entire interval from extreme ultraviolet (UV) to extreme infrared

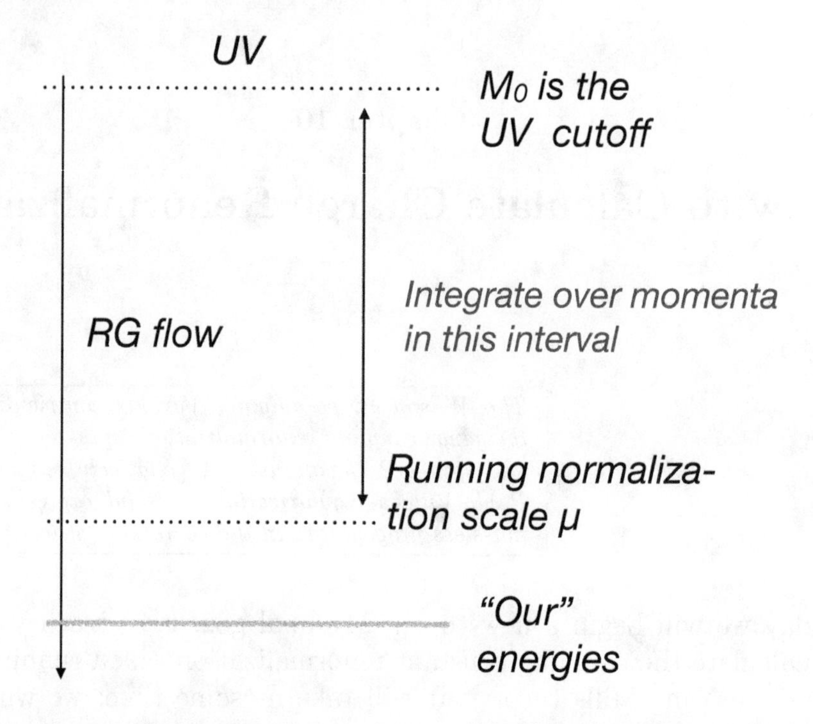

Fig. 10.1 Renormalization group flow. RG evolution from short to larger distances.

(IR), the coupling constants change. Alternatively, people refer to *running* coupling constant. The coupling constant at a momentum scale μ ($M_{\rm IR} \le \mu \le M_{\rm UV}$) is called the running coupling, or, alternatively, the coupling normalized at μ.

In the Abelian QED case we can write

$$S_{\rm UV} = -\frac{1}{4e_0^2}\int dx\, F_{\mu\nu}(x)\, F^{\mu\nu}(x) + \text{matter}\,. \tag{10.1}$$

Here e_0 is the bare charge, to be understood as a charge at the scale $M_0 \equiv M_{\rm UV}$, much larger than any physical scale we could attain. "Much larger" means that all powers of $1/M_0$ can be safely neglected. Still, M_0 is not required to tend to infinity.

The running coupling will be denoted by e^2, without the subscript. In this convention, the covariant derivative in the matter sector is

$$D_\mu = \partial_\mu - iA_\mu\,. \tag{10.2}$$

Then we calculate the contribution to $F_{\mu\nu} F^{\mu\nu}$ coming from loops of matter in the momentum interval limited by M_0 from above and by μ from below. The new coefficient in front of $(-1/4) F_{\mu\nu} F^{\mu\nu}$ will be $[e^2(\mu)]^{-1}$. Our task is to calculate it in terms of the bare coupling and M_0 order by order in perturbation theory. Today we will consider only one-loop correction. *En route* I will teach you some "tricks" that, as a rule, you will not find in other textbooks.

In the old *renormalizability* paradigm it was assumed that M_0 is an auxiliary parameter, and at the end of calculation, one was supposed to take the limit of $M_0 \to \infty$. This gave rise to "infinities" of which theorists were scared. It was Feynman who first noted that physical quantities in QED and similar theories, being expressed in terms of the renormalized coupling constant(s), would have no infinities. In this way "mysterious" infinities were hidden in the relation between the bare and renormalized couplings.

Now we think there are no actual infinities at all. We do not assume that $M_0 \to \infty$. We say that M_0 is the onset of new physics – string theory or something else yet unknown – and introducing $e^2(\mu)$ and expressing quantities at "our" energies in terms of $e^2(\mu)$ also measured at "our" energies is just a way to circumvent our ignorance of so far unknown details of new physics. After all, what can be more natural than this idea? In renormalizable field theories this idea is implementable, and details of new physics at very short distances are irrelevant. In non-renormalizable QFTs they are relevant. As we see, the difference between these two cases is technical rather than conceptual. The difference between the renormalizable and non-renormalizable field theories is that in the latter quantum physics at low energies essentially depends on details of an unknown theory at M_0. Our ignorance of this unknown theory cannot be hidden in a few running coupling constants, as in QED or QCD.

This is a new ideology. Sometimes people *speculate* about effects of the order of $1/M_0^k$ where k is a positive integer to consider possible phenomena highly suppressed at "our" energies.

10.1 Scalar field

This is the only contribution to be found in the "old-fashioned" way. The Lagrangian of the charged (complex) scalar field is

$$\mathcal{L} = (D_\mu \phi)^\dagger (D^\mu \phi) , \qquad D_\mu = \partial_\mu - i A_\mu . \tag{10.3}$$

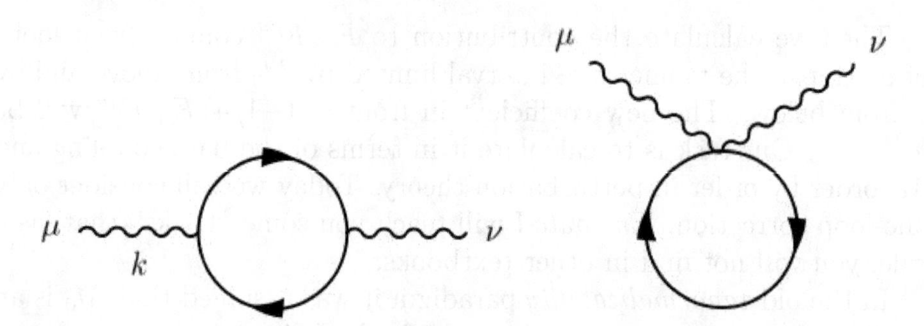

Fig. 10.2 Calculation of the electric charge renormalization in scalar QED at one-loop level. Two Feynman diagrams describe $O(A_\mu A_\nu)$ terms in the effective Lagrangian. The graph on the right is called tadpole. Its only role is to restore gauge invariance in the answer, i.e. the kinematic structure presented in (10.27).

The one-loop contribution is depicted in Fig. 10.2(a). One can write this graph's contribution to the effective Lagrangian $\mathcal{L}_{\text{one-loop}}$ as

$$\frac{-i}{2} A_\mu \int \frac{d^4p}{(2\pi)^4} (2p-q)_\mu (2p-q)_\nu \frac{1}{p^2 - m^2} \frac{1}{(p-q)^2 - m^2} A_\nu \qquad (10.4)$$

The overall result for $\langle j_\mu(q)\, j_\nu(-q)\rangle$ must be transversal (here j_μ is the electromagnetic current in scalar QED). This result is presented by the integral in (10.4). Let us examine it for transversality (hereafter I will omit m^2 which is irrelevant for the question at hand but complicates all expressions),

$$\begin{aligned} & q^\mu \times \int \frac{d^4p}{(2\pi)^4} (2p-q)_\mu (2p-q)_\nu \frac{1}{p^2} \frac{1}{(p-q)^2} \\ &= \int \frac{d^4p}{(2\pi)^4} \left[p^\mu - (p-q)^\mu\right] \left[p^\mu + (p-q)^\mu\right] (2p-q)_\nu \frac{1}{p^2} \frac{1}{(p-q)^2} \\ &= \int \frac{d^4p}{(2\pi)^4} \left[p^2 - (p-q)^2\right] (2p-q)_\nu \frac{1}{p^2} \frac{1}{(p-q)^2} \\ &= \int \frac{d^4p}{(2\pi)^4} (2p-q)_\nu \left[\frac{1}{(p-q)^2} - \frac{1}{p^2}\right]. \end{aligned} \qquad (10.5)$$

Were the last integral in (10.5) well defined, the change of the integration variable in the integrand, $p \to p - q$, would lead us to the conclusion that it vanishes.[1] However, it does not vanish. To see that this is the case, let us assume that the external momentum q is small, and expand the integrand keeping the terms $O(q)$. Then

$$q^\mu \times \int \frac{d^4p}{(2\pi)^4}\,(2p-q)_\mu(2p-q)_\nu\,\frac{1}{p^2}\,\frac{1}{(p-q)^2}$$

$$= \int \frac{d^4p}{(2\pi)^4}\,2p_\nu\,\frac{2pq}{(p)^4} + O(q^2) = \int \frac{d^4p}{(2\pi)^4}\,\frac{q_\nu}{p^2}$$

$$= \frac{-i}{16\pi^2}\,q_\nu\,M_0^2\,. \tag{10.6}$$

Thus, the result is not transversal![2]

In fact, this is not surprising. So far, we have ignored the graph in Fig. 10.2(b) which, as was mentioned, is q independent and has the structure $g_{\mu\nu}M_0^2$ where M_0^2 appears on dimensional grounds. Such diagrams are referred to as *tadpoles*. Adding it to the graph, Fig. 10.2(a) cancels non-transversality.

After these preliminary remarks we can readily calculate the integral in (10.4). On general grounds (Lorentz symmetry) one can write

$$\int \frac{d^4p}{(2\pi)^4}\,(2p-q)_\mu(2p-q)_\nu\,\frac{1}{p^2}\,\frac{1}{(p-q)^2} = \mathcal{A}g_{\mu\nu} + \mathcal{B}q_\mu q_\nu \tag{10.7}$$

where $\mathcal{A}$ and $\mathcal{B}$ depend only on q^2. Now, to get rid of the Lorentz indices we can either convolute μ and ν, or multiply the integral by $q^\mu q^\nu$. In the first case, we arrive at

$$4\mathcal{A} + q^2\mathcal{B} = \int \frac{d^4p}{(2\pi)^4}\,(2p-q)^2\,\frac{1}{p^2}\,\frac{1}{(p-q)^2}\,, \tag{10.8}$$

while in the second case

$$\mathcal{A}q^2 + \mathcal{B}q^4 = \int \frac{d^4p}{(2\pi)^4}\,(2pq-q^2)^2\,\frac{1}{p^2}\,\frac{1}{(p-q)^2} = 0\,. \tag{10.9}$$

The right-hand side of (10.9) vanishes up to irrelevant tadpoles as follows from Eq. (10.6).

---

[1] Indeed, the integrand is odd under $p \leftrightarrow p-q$. Shifts of integration variable are legitimate for all convergent integrals and those logarithmically divergent. If the divergence is stronger than logarithmic, such shifts are not allowed, see below.

[2] Since the result is a linear function of q times a q-independent tadpole, it is valid not only for small values of q, but for any value.

Equation (10.9) implies $\mathcal{B}q^2 = -\mathcal{A}$ and then Eq. (10.8) takes the form

$$3\mathcal{A} = \int \frac{d^4p}{(2\pi)^4} \left\{2\left[p^2\right] + 2\left[p^2 - 2pq + q^2\right] - q^2\right\} \frac{1}{p^2} \frac{1}{(p-q)^2}, \quad (10.10)$$

where I rearranged the numerator in (10.8) in a special way. Namely, the first square bracket cancels the first denominator, leaving us with a tadpole (to be discarded) while the second square bracket cancels the second denominator. This tadpole also must be discarded. As a result, we obtain

$$\mathcal{A} = -\frac{q^2}{3} \int \frac{d^4p}{(2\pi)^4} \frac{1}{p^2} \frac{1}{(p-q)^2}. \quad (10.11)$$

A simple examination shows that the above integral logarithmically diverges at large momenta (i.e. in the ultraviolet, UV), and we need to regularize it in order for this integral to make sense. The simplest and the oldest method of regularization belongs to Pauli and Villars (PV). In the next subsection, I review this method which is perfectly good at one loop. Meanwhile, assembling Eqs. (10.4), (10.7), and (10.11) we can present the expression for $\mathcal{L}_{\text{one-loop}}$ as follows:

$$\mathcal{L}_{\text{one-loop}} = \frac{-i}{2} A^\mu \left\{\frac{\mathcal{A}}{q^2}\left(q^2 g_{\mu\nu} - q_\mu q_\nu\right)\right\} A^\nu \quad (10.12)$$

10.2 Pauli–Villars regularization

For pedagogical purposes, I we will calculate the integral using the PV regularization. For dimensional regularization which is very popular at present, see pages 249-251 in Peskin and Schroeder. The original integral in (10.11) is replaced by

$$\mathcal{I} = \int \frac{d^4p}{(2\pi)^4} \left\{\frac{1}{p^2} \frac{1}{(p-q)^2} - \frac{1}{\left(p^2 - M_0^2\right)^2}\right\} \quad (10.13)$$

where the second term on the right-hand side is due to the PV loop which is subtracted from the "physical" loop. The PV parameter M_0 is assumed to be much larger than any physical parameter of dimension of mass; therefore, q can be neglected in the PV term. As I explained in QFT I course, it is convenient to perform the Wick rotation in the Euclidean space, $p_0 \to ip_0$, (where p_0 is the time component of the loop momentum) and, correspondingly, $p^2 \to -p^2$ in the Euclidean. If q^2 is negative – and we will assume this – one can always choose a reference frame in which the time component $q_0 = 0$, hence $2pq = -2\vec{p}\vec{q} \to -2pq$ (the latter expression, $-2pq$, assumes that we are in the Euclidean space). Simultaneously q^2 becomes $-q^2$ in the Euclidean.

Keeping in mind that the four-dimensional solid angle is

$$\Omega_4 = 2\pi^2 \tag{10.14}$$

we can transform (10.13) in the Euclidean space into

$$\mathcal{I} = i \int \frac{dp^2 \cdot p^2}{16\pi^2} \int_0^1 dx \left\{ \left[\frac{1}{p^2 + x(1-x)(-q^2)} \right]^2 - \left[\frac{1}{p^2 + M_0^2} \right]^2 \right\}$$

$$= \frac{i}{16\pi^2} \int_0^1 dx \log \frac{M_0^2}{x(1-x)(-q^2)} \tag{10.15}$$

Above we used the Feynman parametrization. Moreover,

$$\mathcal{I} = \frac{i}{16\pi^2} \left\{ \log \frac{M_0^2}{(-q^2)} + 2 \right\}, \quad \mathcal{A} = -\frac{q^2}{3} \mathcal{I}. \tag{10.16}$$

We arrive at the final result for the scalar loop contribution to $\mathcal{L}_{\text{one-loop}}$,

$$\mathcal{L}_{\text{one-loop}} = -\frac{1}{12} \frac{1}{16\pi^2} \left\{ \log \frac{M_0^2}{(-q^2)} + 2 \right\} F_{\mu\nu} F^{\mu\nu}, \tag{10.17}$$

to be compared with the "bare" action in UV,

$$\mathcal{L} = -\frac{1}{4e_0^2} F_{\mu\nu} F^{\mu\nu}. \tag{10.18}$$

I would like to pause here to note that the coefficient in front of the logarithm is uniquely determined and will be the same in any method of regularization. However, in principle, other methods can produce in (10.17) a different additive constant. This additive constant is equivalent to replacing the upper cut-off $M_0^2 \to CM_0^2$, and then [3]

$$\log \left(\frac{M_0^2}{m^2} \right) \to \log \left(\frac{CM_0^2}{m^2} \right) \to \log \left(\frac{M_0^2}{m^2} \right) + \log C, \tag{10.19}$$

where C is a pure number. Your final physical amplitude should be C independent after renormalization. This means that if a regularization/renormalization procedure is chosen, you have to stick to it till the very end. In what follows we will need only the universal logarithmic term.

[3]See Problem 18 on page 264.

If one wants to obtain only the logarithmic term at one loop (i.e. ignore possible additive constants), this goal can be achieved even faster without applying a careful regularization – the Pauli–Villars one or any other. Indeed, suffice it to write the integral (10.15) as follows:

$$\mathcal{I} = i \int_{\mu^2}^{M_0^2} \frac{dp^2 \cdot p^2}{16\pi^2} \left(\frac{1}{p^2}\right)^2 = \frac{i}{16\pi^2}\left[\log\left(\frac{M_0^2}{\mu^2}\right)\right], \tag{10.20}$$

where I introduced a rigid *ad hoc* cut off in the UV (M_0^2) and a rigid cut off in the infrared (IR), denoting the infrared cut off by μ^2. The coefficient of $\log M_0^2$ is not affected by crudeness of the above procedure and remains the same as under a more subtle Pauli-Villars regularization (cf. (10.16) and (10.20)). I will use this simplifying knowledge in calculating (10.4).

10.3 Calculation of the scalar field loop

Let us return to Eq. (10.4). Since we are interested only in the coefficient in front of the logarithm, we will apply a crude rigid cut off, which means that the mass terms in Eq. (10.4) can be omitted, and it takes the form

$$\begin{aligned} &A^\mu \int \frac{d^4p}{(2\pi)^4}\,(2p-q)_\mu(2p-q)_\nu\,\frac{1}{p^2}\,\frac{1}{(p-q)^2}\,A^\nu \\ &= A_\mu \int \frac{d^4p}{(2\pi)^4}\,(4p_\mu p_\nu + q_\mu q_\nu - 2q_\mu p_\nu - 2q_\nu p_\mu)\,\frac{1}{p^2}\,\frac{1}{(p-q)^2}\,A_\nu\,. \end{aligned} \tag{10.21}$$

Keeping in mind that QED is gauge invariant (and that the tadpole q-independent graph of Fig. 10.2(b) has to be added for maintaining the gauge invariance), and the dimensions of the integral and integrand, it is easy to see that the final result of calculation of (10.21) must have the form

$$A^\mu\left(g_{\mu\nu}q^2 - q_\mu\, q_\nu\right)A^\nu \times \text{const} \times \left[\log\left(\frac{M_0^2}{\mu^2}\right) + \text{const}'\right]. \tag{10.22}$$

I remind that we are not interested in const$'$. The transverse q structure in (10.22) is the only one which remains invariant under the gauge transformation $A^\mu \to A^\mu + \alpha q^\mu$. It is important that it is quadratic in q. Terms $O(q^0)$ cannot be gauge invariant and will be killed by the tadpole. On dimensional grounds terms $O(q^4)$ cannot contain logarithms. Hence we will focus only on quadratic in q terms in (10.21). To extract them we need to expand the denominator in (10.21) assuming that $q \ll p$,

$$\frac{1}{(p-q)^2} \to \frac{1}{p^2\left(1-\frac{2pq}{p^2}+\frac{q^2}{p^2}\right)} = \frac{1}{p^2}\left[1+\frac{2pq}{p^2}-\frac{q^2}{p^2}+\frac{4(pq)^2}{p^4}+...\right]. \tag{10.23}$$

Now we can combine (10.23) and (10.21) and perform multiplication keeping only $O(q^2)$ terms. Since the denominator will now contain only powers of p^2, we can average over the solid angle using the rules

$$\overline{p_\mu p_\nu} = \frac{1}{4} g_{\mu\nu}\, p^2\,, \qquad \overline{p_\mu p_\nu p_\alpha p_\beta} = \frac{1}{24}\left(g_{\mu\nu}g_{\alpha\beta} + g_{\mu\alpha}g_{\nu\beta} + g_{\mu\beta}g_{\nu\alpha}\right)p^4\,. \tag{10.24}$$

The standard Wick rotation

$$p_0 \to i p_0\,, \qquad p^2 \to -p^2 \tag{10.25}$$

leads us, after assembling all terms, to the following result for (10.4):

$$A^\mu \int_{\mu^2}^{M_0^2} i\,\frac{dp^2 \cdot p^2}{16\pi^2}\left(\frac{1}{p^4}\right)\left(-\frac{1}{3}\right)\left(g_{\mu\nu}q^2 - q_\mu\, q_\nu\right)A^\nu\,. \tag{10.26}$$

Please, observe the emergence of the transverse q structure. It appears automatically once we ignore terms $O(q^0)$.

Now, let us restore the overall factors e and i omitted so far. In the normalization we currently use, all relevant vertices do not contain e factors. Neither these factors appear in the matter field propagators. Therefore, our final result will be $O(e^0)$, to be compared with Eq. (10.1).

As to the i factors, they are as follows: the amplitude we calculated contains overall i because of e^{iS}, i^2 come from two propagators, another i^2 from the vertices (see Fig. 10.2(a)) and $(-i)^2$ from the definition of the covariant derivative in Eq. (10.3). Combining all these factors with the factor i from (10.26) we get for Eq. (10.4)

$$-\frac{1}{3}\,\frac{1}{16\pi^2}\log\left(\frac{M_0^2}{\mu^2}\right)A^\mu\left(g_{\mu\nu}q^2 - q_\mu\, q_\nu\right)A^\nu\,. \tag{10.27}$$

This is not quite the end of the story, however. Equation (10.27) is the amplitude. We want to compare it to the action (10.3). In passing from the amplitude to the action we must divide by two because of combinatorics and also take into account that

$$F_{\mu\nu}F^{\mu\nu} \leftrightarrow 2\left[q^2A^2 - (qA)^2\right]. \tag{10.28}$$

As a result, we arrive at

$$S_0 + \Delta S = -\frac{1}{4e_0^2} \int dx\, F_{\mu\nu}(x)\, F^{\mu\nu}(x)$$

$$-\frac{1}{48\pi^2} \log\left(\frac{M_0^2}{\mu^2}\right) \frac{1}{4} \int dx\, F_{\mu\nu}(x)\, F^{\mu\nu}(x)\,. \tag{10.29}$$

This one-loop formula implies, in turn, that

$$\frac{1}{e^2} = \frac{1}{e_0^2} + \frac{1}{48\pi^2} \log\left(\frac{M_0^2}{\mu^2}\right), \qquad \frac{1}{e_0^2} = \frac{1}{e^2} - \frac{1}{48\pi^2} \log\left(\frac{M_0^2}{\mu^2}\right) \tag{10.30}$$

or, equivalently,

$$\alpha(\mu) = \frac{\alpha_0}{1 + \frac{1}{12\pi}\,\alpha_0 \log \frac{M_0^2}{\mu^2}}\,, \qquad \alpha_0 = \frac{\alpha(\mu)}{1 - \frac{1}{12\pi}\,\alpha(\mu) \log \frac{M_0^2}{\mu^2}}\,. \tag{10.31}$$

Certainly, only $O(\alpha^2)$ term on the right-hand side of (10.31) presents a legitimate calculation of one loop; higher order terms are not given by one loop. Later we will study renormalization group, and will learn how to perform summation of logarithms.

10.4 Calculation of the fermion loop

Now, when we have gained some experience, the calculation of the fermion field contribution will not take us much time. Note that for the fermion loop the tadpole graph of Fig. 10.2(b) is absent. The only diagram to be considered is that of Fig. 10.2(a).

The fermion analog of Eq. (10.4) is

$$(-1) \int \frac{d^4p}{(2\pi)^4} \mathrm{Tr}\left[\gamma_\mu \frac{\gamma^\alpha\, p_\alpha}{p^2 - m^2}\, \gamma^\nu \frac{\gamma^\beta\, (p-q)_\beta}{(p-q)^2 - m^2}\right]. \tag{10.32}$$

The (-1) factor is due to the fact that fermions, not bosons, propagate in the loop. The operational strategy is the same as in the boson case. After we single out terms $O(q^2)$ in the integrand, as we know, the result will be automatically transversal, i.e. proportional to $(g_{\mu\nu}q^2 - q_\mu\, q_\nu)$. Being aware of this circumstance, we can convolute μ and ν from the very beginning, then

$$(g_{\mu\nu}q^2 - q_\mu\, q_\nu) \to 3q^2\,. \tag{10.33}$$

Let us now perform the convolution directly in (10.32). Then we obtain

$$8 \int \frac{d^4p}{(2\pi)^4} \left[\frac{p^2 - pq}{p^2}\, \frac{1}{(p-q)^2}\right]. \tag{10.34}$$

Next, use the identity

$$p^2 - pq = \frac{1}{2}(p-q)^2 + \frac{p^2}{2} - \frac{q^2}{2} \tag{10.35}$$

The first term in (10.35) will cancel the second denominator in (10.34), leaving us with a tadpole-like contribution. The second term in (10.35) will cancel the first denominator in (10.34), again leaving us with a tadpole-like contribution. Both can be discarded.

What remains is

$$-4\,q^2 \int \frac{d^4p}{(2\pi)^4}\left[\frac{1}{p^2}\,\frac{1}{(p-q)^2}\right] \to -4\,q^2 \int \frac{d^4p}{(2\pi)^4}\,\frac{1}{p^4} + \ldots \tag{10.36}$$

The integral in (10.36) has already been calculated, see Sec. 10.2. Using (10.33) we reconstruct the full answer for the amplitude,

$$\left(-\frac{4}{3}\right)\frac{i}{16\pi^2}\left[\log\left(\frac{M_0^2}{\mu^2}\right)\right] A^\mu\left(g_{\mu\nu}q^2 - q_\mu\,q_\nu\right)A^\nu$$

$$\to \left(-\frac{1}{3}\right)\frac{i}{16\pi^2}\left[\log\left(\frac{M_0^2}{\mu^2}\right)\right] F_{\mu\nu}\,F^{\mu\nu}\,, \tag{10.37}$$

where in passing to $F_{\mu\nu}\,F^{\mu\nu}$, I applied the combinatorial (1/2) factor and (10.28). The i factors can be counted as follows: the amplitude we calculated must additionally contain overall i because of e^{iS}, and i^2 comes from two propagators. Combining all these factors with the factor i from (10.37) we finally obtain

$$S_0 + \Delta S = -\frac{1}{4e_0^2}\int dx\, F_{\mu\nu}(x)\,F^{\mu\nu}(x)$$

$$-\frac{1}{12\pi^2}\log\left(\frac{M_0^2}{\mu^2}\right)\frac{1}{4}\int dx\, F_{\mu\nu}(x)\,F^{\mu\nu}(x)\,. \tag{10.38}$$

The sign of the correction is the same as in (10.29) and the value of the coefficient in front of $F_{\mu\nu}(x)\,F^{\mu\nu}(x)$ is four times larger. This was to be expected from the calculations we had carried out in the spring semester.

The analogs of Eqs. (10.30) and (10.31) are

$$\frac{1}{e^2} = \frac{1}{e_0^2} + \frac{1}{12\pi^2}\log\left(\frac{M_0^2}{\mu^2}\right), \qquad \frac{1}{e_0^2} = \frac{1}{e^2} - \frac{1}{12\pi^2}\log\left(\frac{M_0^2}{\mu^2}\right) \tag{10.39}$$

or, equivalently,

$$\alpha(\mu) = \frac{\alpha_0}{1 + \frac{1}{3\pi}\,\alpha_0 \log\frac{M_0^2}{\mu^2}}, \qquad \alpha_0 = \frac{\alpha(\mu)}{1 - \frac{1}{3\pi}\,\alpha(\mu)\log\frac{M_0^2}{\mu^2}}. \tag{10.40}$$

Note that in both cases above, the charge at the normalization point μ (at a lower Euclidean momentum than M_0) is smaller than the bare charge, and the other way around, the bare charge is larger.

This phenomenon (known as screening) is general for matter field loops and has a physically transparent interpretation. See figure on p. 255 of Peskin and Schroeder and Appendix 14.1.

Caveat: the above results were obtained keeping logarithms and neglecting additive constants. Our method of cut off is too crude to carefully treat momenta of the order of the particle mass. Therefore, it must be assumed that

$$m \ll \mu \ll M_0 \,. \tag{10.41}$$

Appendix 10.1. A Very Useful Appendix [4]

I will now teach you a method of calculating loop diagrams of the type depicted in Fig. 10.2(a) which will minimize your effort. It is applicable provided that all mass terms in the loop propagators can be considered negligible and can be ignored. The basic idea is simple: for massless fields propagators are as simple in the coordinate space as they are in the momentum space, while your calculation will (formally) "loose" one loop compared to that in the momentum.

This method is applicable not only in QCD with light quarks, but also in conformal field theories, in the Schwinger model (Problem 27a* on page 272), in the Glashow-Weinberg-Salam model, etc.

The price you will have to pay is the universal formula for the Fourier transformations. Let us start from four dimensions and define

$$F(p) = \int d^4x \, e^{ipx} \, F(x) \,, \qquad F(x) = \int \frac{1}{(2\pi)^4} \, e^{-ipx} \, F(p) \,. \tag{10.42}$$

The most frequently used transformations are

$$\frac{1}{p^2} \leftrightarrow \frac{i}{4\pi^2} \frac{1}{x^2} \,, \tag{10.43}$$

$$\frac{p^\mu}{p^2} \leftrightarrow \frac{1}{2\pi^2} \frac{x^\mu}{x^4} \,, \tag{10.44}$$

$$\frac{p^\mu}{p^4} \leftrightarrow \frac{1}{8\pi^2} \frac{x^\mu}{x^2} \,. \tag{10.45}$$

[4] This Appendix is based on [1; 2].

All the above expressions can be derived as follows. If we define $A_D(p^2)$ in D dimensions as

$$A_D(p^2) = \int d^D x\, e^{ipx} \frac{1}{(x^2)^n} \xrightarrow{\text{Euclid}} (-i)(-1)^n \int_{\text{Euclid}} d^D x\, e^{-ipx} \frac{1}{(x^2)^n}\,, \tag{10.46}$$

where we have passed to the Euclidean space. Integrating over angles, results in

$$A_D(p^2) = -i\,(-1)^n\, 2^{D/2}\, \pi^{D/2}\, |p|^{1-\frac{1}{2}D}$$

$$\times \int d|x|\, |x|^{\frac{1}{2}D-2n}\, J_{\frac{1}{2}D-1}(|p|\,|x|)$$

$$= -i\,(-1)^n\, 2^{D-2n}\, \pi^{\frac{1}{2}D}\, |p|^{2n-D} \frac{\Gamma\left(\frac{D}{2}-n\right)}{\Gamma(n)}\,, \tag{10.47}$$

where J is the corresponding Bessel function (with the index $\frac{1}{2}D-1$). If $D \to 4$ this expression can be singular (e.g. at $n=2$, it has a pole). However, the corresponding residue is proportional to $(p^2)^{n-2}$. If we are interested only in logarithmic terms, polynomials are inessential. They may be needed, however, if you need to carry out renormalization program with non-logarithmic accuracy.

You write $D = 4-\epsilon$ and perform the ϵ expansion. The nonsingular part of (10.47) *after we return to the Minkowski space* has the form

$$\int d^4 x \frac{1}{(x^2)^n}\, e^{ipx} = \frac{i\,(-1)^n\, 2^{4-2n}\, \pi^2}{\Gamma(n-1)\,\Gamma(n)}\, (p^2)^{n-2} \log(-p^2)\,. \tag{10.48}$$

To obtain Eqs. (10.43)–(10.45) one should return to (10.47). For $n=1$ and $d=4$ this expression is nonsingular and returning to the Minkowski space we reproduce (10.43). Differentiating $\int d^4x\, e^{ipx}\,(x^2)^{-1}$ over p_μ we arrive at (10.45) starting from (10.43). Equation (10.44) can be obtained from (10.48) by setting $n=2$ and differentiating both sides of (10.48) over p_μ.

Equation (10.47) (written in Euclidean space) is valid for arbitrary values of D. Above it was used in the $D=4$ case. In two dimensions (i.e. $D=2$) Eq. (10.47) stays intact with the substitution $D=2$ or, if needed, $D=2-\epsilon$. In the second equation in (10.42), one must replace $(2\pi)^4$ in the denominator by $(2\pi)^2$ or $(2\pi)^D$.

In this way in two dimensions we arrive at

$$\frac{1}{x^2} \leftrightarrow -i\pi \log p^2 + \text{const}\,, \tag{10.49}$$

$$\frac{x^\mu}{x^2} \leftrightarrow -2\pi\,\frac{p^\mu}{p^2}\,. \tag{10.50}$$

Another useful formula is

$$\int d^D x\, e^{ipx}\,\frac{x^\mu\, x^\nu}{x^4} = i\,2^{D-3}\,\pi^{\frac{1}{2}D}\,\Gamma\left(\frac{D}{2}-1\right)(-p^2)^{1-\frac{1}{2}D}$$
$$\times\left[g^{\mu\nu} + 2\left(1-\frac{D}{2}\right)\frac{p^\mu\,p^\nu}{p^2}\right]. \tag{10.51}$$

*****

As an exercise let us calculate the four-dimensional fermion loop shown in Fig. 10.2(a) (assuming one Dirac fermion of charge 1). (For a similar calculation in two dimensions see Problem 27a* on page 272.) We will need to use Eqs. (10.46) and (10.47) with $D = 4$ and $n = 3$.

$$\Pi^{\mu\nu}(x) = ie^2\left\langle \bar\psi(x)\gamma^\mu\psi(x)\,,\bar\psi(0)\gamma^\nu\psi(0)\right\rangle$$
$$= ie^2\,\mathrm{Tr}\left(\gamma^\mu\,\frac{i}{2\pi^{\frac{1}{2}D}}\,\frac{x_\alpha\,\gamma^\alpha}{(x^2)^{\frac{1}{2}D}}\,\gamma^\nu\,\frac{i}{2\pi^{\frac{1}{2}D}}\,\frac{x_\beta\,\gamma^\beta}{(x^2)^{\frac{1}{2}D}}\right)$$
$$= -ie^2\,\frac{1}{4\pi^D}\,\frac{1}{(x^2)^D}\left(2x^\mu\,x^\nu - g^{\mu\nu}x^2\right). \tag{10.52}$$

The first thing to do is to check transversality,

$$\partial_\mu\left[\frac{1}{(x^2)^D}\left(2x^\mu\,x^\nu - g^{\mu\nu}x^2\right)\right]$$
$$= \frac{2D\,x^\nu}{(x^2)^D} - \frac{2D\,x^\mu}{(x^2)^{D+1}}\left(2x^\mu\,x^\nu - g^{\mu\nu}x^2\right) = 0\,. \tag{10.53}$$

For $D = 4$, the concluding calculation is quite trivial. Since we have already verified the transversality we can convolute μ and ν and then we arrive at

$$\Pi^\mu_\mu(x) = i\,e^2\frac{1}{2\pi^4}\,\frac{1}{(x^2)^3}\,,\qquad \Pi^\mu_\mu(p^2) = e^2\,\frac{1}{16\pi^2}\,p^2\log(-p^2)\,, \tag{10.54}$$

which, in turn, implies

$$\Pi^{\mu\nu}(p^2) = e^2 \frac{1}{48\pi^2} \left(g^{\mu\nu} p^2 - p^\mu p^\nu\right) \log(-p^2)\,, \tag{10.55}$$

cf. Sec. 10.4.

References

[1] *Vacuum Structure and QCD Sum Rules*, Ed. M. Shifman, (North Holland, Amsterdam, 1992), p. 259.
[2] M. Shifman, *ITEP Lectures on Particle Physics and Field Theory*, (World Scientific, Singapore, 1999), Vol. 2, p. 707.

Chapter 11

Lecture 11. Yang-Mills Coupling (*Continued*)

Yang-Mills theory with matter. Renormalizing the gauge coupling at one loop using the background field method. Rederiving appropriate Faddeev-Popov ghosts. Calculation of the ghost loop and its role. Asymptotic freedom.

11.1 The scalar and fermion field loops

The scalar field result in QED was

$$S_0 + \Delta S = -\frac{1}{4e_0^2}\int dx\, F_{\mu\nu}(x)\, F^{\mu\nu}(x)$$

$$-\frac{1}{48\pi^2}\log\left(\frac{M_0^2}{\mu^2}\right)\frac{1}{4}\int dx\, F_{\mu\nu}(x)\, F^{\mu\nu}(x)\,. \tag{11.1}$$

This one-loop formula implies, in turn, that

$$\frac{1}{e^2} = \frac{1}{e_0^2} + \frac{1}{48\pi^2}\log\left(\frac{M_0^2}{\mu^2}\right), \qquad \frac{1}{e_0^2} = \frac{1}{e^2} - \frac{1}{48\pi^2}\log\left(\frac{M_0^2}{\mu^2}\right) \tag{11.2}$$

or, equivalently,

$$\alpha(\mu) = \frac{\alpha_0}{1+\frac{1}{12\pi}\alpha_0\log\frac{M_0^2}{\mu^2}}, \qquad \alpha_0 = \frac{\alpha(\mu)}{1-\frac{1}{12\pi}\alpha(\mu)\log\frac{M_0^2}{\mu^2}}\,. \tag{11.3}$$

The Dirac spinor field result in QED was

$$S_0 + \Delta S = -\frac{1}{4e_0^2}\int dx\, F_{\mu\nu}(x)\, F^{\mu\nu}(x)$$

$$-\frac{1}{12\pi^2}\log\left(\frac{M_0^2}{\mu^2}\right)\frac{1}{4}\int dx\, F_{\mu\nu}(x)\, F^{\mu\nu}(x)\,. \tag{11.4}$$

The sign of the correction is the same as in (11.1) and the value of the coefficient in front of $F_{\mu\nu}(x)\, F^{\mu\nu}(x)$ is four times larger. This was to be expected from the calculations we had carried out in the previous (spring) semester.

The only modifications needed in order to convert these results into Yang-Mills theory is the generator matrices in the matter vertices following from the covariant derivative. In the non-Abelian case

$$D_\mu = \partial_\mu - iA_\mu^a T^a\,, \qquad F_{\mu\nu} \to G_{\mu\nu}^a = \partial_\mu A_\nu^a - \partial_\nu A_\mu^a + f^{abc} A_\mu^b A_\nu^c\,. \quad (11.5)$$

The properties of the generators are summarized on page 83. The non-Abelian charge renormalization is described by the same graph as in Fig. 10.2(a) with the replacement of the external photon legs by the gluon legs. The requirement of the non-Abelian gauge invariance implies that the one-loop contribution is fully determined by Fig. 11.1, with just two A_μ^a attached. The terms $O(A^3)$ and $O(A^4)$ needed to obtain the full operator $G_{\mu\nu}^2$ will follow from the gauge invariance.

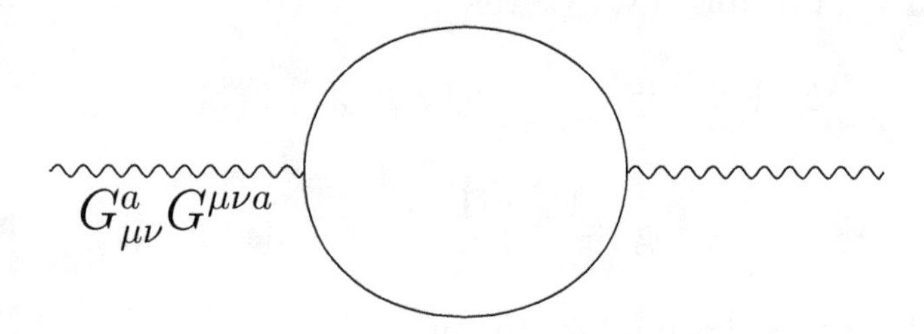

Fig. 11.1 The solid line loop: (a) complex scalar field in the fundamental representation; (b) Dirac fermion field in the fundamental representation; (c) gluon field and ghost field in the adjoint representation. Wavy lines represent the gauge field A_μ. This graph shows two external gluon legs. The cubic and quartic terms in $(G_{\mu\nu a}\, G^{\mu\nu\, a})_{\rm ext}$ automatically follow from gauge invariance.

Hence, assuming the matter fields are in the fundamental representation the only change in the previous calculation is the emergence of the extra

$$\mathrm{Tr}\, T^a T^b = \frac{1}{2}\,\delta^{ab} \qquad (11.6)$$

(see Appendix 9.1).

Equations (11.1) and (11.4) take the form

$$S_0 + \Delta S = -\frac{1}{4g_0^2}\int dx\, G^a_{\mu\nu}(x)\, G^{\mu\nu\, a}(x)$$

$$-\frac{1}{96\pi^2}\log\left(\frac{M_0^2}{\mu^2}\right)\frac{1}{4}\int dx\, G^a_{\mu\nu}(x)\, G^{\mu\nu\, a}(x)\,.$$

$$\text{for scalars}\,, \tag{11.7}$$

and

$$S_0 + \Delta S = -\frac{1}{4g_0^2}\int dx\, G^a_{\mu\nu}(x)\, G^{\mu\nu\, a}(x)$$

$$-\frac{1}{24\pi^2}\log\left(\frac{M_0^2}{\mu^2}\right)\frac{1}{4}\int dx\, G^a_{\mu\nu}(x)\, G^{\mu\nu\, a}(x)\,.$$

$$\text{for spinors}\,. \tag{11.8}$$

11.2 The gauge field and ghost loops

Now I will teach you how to calculate the gauge field and ghost contributions in charge renormalization in a simpler way than in Peskin and Schroeder or other textbooks. Let us start from the gauge fields. In the diagram depicted in Fig. 11.1 the solid line represents, in particular, the gauge field propagation. The wavy lines can be interpreted as an "external" gauge field.

To separate the external field from the quantum field propagating in the loop we can always write

$$A^a_\mu \equiv \left(A^a_\mu\right)_{\text{ext}} + a^a_\mu\,, \tag{11.9}$$

where a^a_μ is the quantum part (the field which propagates in the loop of Fig. 11.1). Substituting (11.9) in the Lagrangian (1.33) and keeping only the terms quadratic in a^a_μ (since all other powers of a^a_μ are irrelevant for the graph in Fig. 11.1) we obtain

$$\mathcal{L} = \frac{1}{g_0^2}\left\{-\frac{1}{4}\left(G^a_{\mu\nu}G^{\mu\nu\, a}\right)_{\text{ext}} - \frac{1}{2}\left(D^{\text{ext}}_\mu a^a_\nu\right)^2 + \frac{1}{2}\left(D^{\text{ext}}_\mu a^a_\nu\right)\left(D^{\text{ext}}_\nu a^a_\mu\right)\right.$$

$$\left. + \frac{1}{2}a^a_\mu \left(G^b_{\mu\nu}\right)_{\text{ext}} f^{abc} a^c_\nu\right\} + \ldots \tag{11.10}$$

For our purposes the general gauge transformation (1.28) can be represented as (at small ω)

$$\left(A_\mu^a\right)_{\rm ext} \to \left(A_\mu^a\right)_{\rm ext}\,, \tag{11.11}$$

$$a_\mu^a \to a_\mu^a + D_\mu^{\rm ext}\omega^a\,. \tag{11.12}$$

where [1]

$$D_\mu^{\rm ext}\omega^a(x) \equiv \partial_\mu\,\omega^a - i\left(A_\mu^b\right)_{\rm ext}\left(T^b\right)^{ac}\omega^c \equiv \partial_\mu\,\omega^a + f^{abc}\left(A_\mu^b\right)_{\rm ext}\omega^c\,. \tag{11.13}$$

One can check (and I recommend this verification as a home exercise) that the Lagrangian (11.10) is invariant under (11.11) and (11.12).

Naturally, the action of $D_\mu^{\rm ext}$ on a_μ is defined as

$$D_\mu^{\rm ext}a_\nu^a(x) \equiv \partial_\mu\,a_\nu^a - i\left(A_\mu^b\right)_{\rm ext}\left(T^b\right)^{ac}a_\nu^c \equiv \partial_\mu\,a_\nu^a + f^{abc}\left(A_\mu^b\right)_{\rm ext}a_\nu^c\,. \tag{11.14}$$

Now, here is a smart trick. Let us impose the gauge condition in the form

$$D_\mu^{\rm ext}a_\mu^a = 0\,, \qquad G(a(x)) = D_\mu^{\rm ext}a_\mu^a(x) \tag{11.15}$$

rather than the standard $\partial_\mu A_\mu^a = 0$, see e.g. Chapter 8, page 74. Then the determinant of $\delta G\left(a^\omega(x)\right)/\delta\omega(y)$ becomes

$$\det\frac{\delta G\left(a^\omega(x)\right)}{\delta\omega(y)} = D_\mu^{\rm ext}D^{\mu\,\rm ext}\,\delta(x-y)\,, \tag{11.16}$$

cf. Eqs. (8.18) and (9.21). The derivatives on the right-hand side enter symmetrically!

Unlike the Abelian case, this determinant cannot be dropped. It can be represented in terms of the ghost fields $\bar c$, c contributing in the Lagrangian as follows:

$$\mathcal{L}_{\rm ghost} = \left(D_\mu^{\rm ext}\bar c\right)\left(D^{\mu\,\rm ext}c\right)\,. \tag{11.17}$$

The ghost fields above are scalar complex fields in the adjoint representation of the gauge group and "wrong" statistics. They appear in the loop in Fig. 11.1 because the ghost term (11.17) contains interaction between the ghost fields $\bar c$, c and the external field $A_\mu^{\rm ext}$ through the covariant derivatives. If you compare (11.17) with Eqs. (9.21) and (9.24) you will see that the symmetry in the ghost term absent in Chapter 9 is restored, which will help us a lot.

[1]The group generators T^a in Eqs. (11.13) and (11.14) must be taken in the adjoint representation.

Following the procedure described at the end of Chapter 8, page 76, we can add to the Lagrangian (11.10) the following gauge fixing term

$$\Delta\mathcal{L}_{\text{gauge}} = -\frac{1}{2g^2}\left(D_\mu^{\text{ext}} a_\mu^a\right)^2 . \tag{11.18}$$

The third term in the first line of (11.10) can be reduced to (11.18) (with the opposite sign) plus the fourth term in (11.10) by performing two integration by parts. In doing so, we must take into account that the covariant derivatives D_μ^{ext} do not commute, namely,

$$[D_\mu^{\text{ext}}, D_\nu^{\text{ext}}] = iG_{\mu\nu}^b T^b . \tag{11.19}$$

The covariant derivative D_ν^{ext} in (11.10) should be dragged just to the left of a_ν in the first bracket and D_μ^{ext} just to the left of a_μ in the second bracket in the expression $\left(D_\mu^{\text{ext}} a_\nu^a\right)\left(D_\nu^{\text{ext}} a_\mu^a\right)$.

After this procedure, the gauge-fixing term (11.18) will cancel, while the last term in (11.10) will double, and finally we will arrive at the following Lagrangian for the quantum part of the gauge field:

$$\mathcal{L}_a = \frac{1}{g_0^2}\left\{-\frac{1}{4}\left(G_{\mu\nu}^a G^{\mu\nu\, a}\right)_{\text{ext}} - \frac{1}{2}\left(D_\mu^{\text{ext}} a_\nu^a\right)^2 + a_\mu^a \left(G_{\mu\nu}^b\right)_{\text{ext}} f^{abc} a_\nu^c\right\} + \ldots \tag{11.20}$$

The first term in (11.20) is for comparison, the second term will give us the a propagator and "charge" vertices, the third term determines the spin coupling of the external (background) field with a quanta. This will play the most crucial role in what follows.

For a short while, let us forget about the spin coupling, and focus on the $-\frac{1}{2}\left(D_\mu^{\text{ext}} a_\nu^a\right)^2$ term. We see that this term is nothing but four replicas of a real scalar field in the adjoint representation.[2] Its contribution to the gauge coupling renormalization is trivially inferred from Eq. (11.7).

The Green's function for a_μ is

$$\left\langle a_\mu^a\, a_\nu^b\right\rangle \to \frac{\delta_{\mu\nu}\delta^{ab}\, i}{k^2}\, g_0^2 . \tag{11.21}$$

The vertex $aA^{\text{ext}}a$ is the same as for the scalar field with the replacement of the fundamental representation generator by the adjoint representation

$$\left(T^b\right)_{ac} = i f^{abc} . \tag{11.22}$$

[2]The $\nu = 0$ scalar field has a wrong sign in the kinetic energy. This becomes unimportant after Euclidean rotation, see also (11.21).

Note that the trace in the adjoint is

$$\mathrm{Tr}\,(T^b T^{\tilde b})_{\mathrm{adj}} = f^{abc} f^{a\tilde b c} = N\delta^{b\tilde b}, \quad \text{cf.} \quad \mathrm{Tr}\,(T^b T^{\tilde b})_{\mathrm{fund}} = \frac{1}{2}\delta^{b\tilde b}. \tag{11.23}$$

Thus, in order to obtain the non-spin gluon contribution to the charge renormalization, we must divide (11.7) by two (proceeding from complex to real field), multiply by $2N$ to take into account the difference in the generator traces, and multiply by 4 (because we deal here with four scalar fields). As a result, we get

$$\Delta S_{g1} = -\frac{N}{24\pi^2}\log\left(\frac{M_0^2}{\mu^2}\right)\frac{1}{4}\int dx\, G^a_{\mu\nu}(x)\, G^{\mu\nu\, a}(x)$$

$$\text{for non-spin part of the gauge field}\,. \tag{11.24}$$

Let us deal with the ghost contribution in the same manner. There are two distinctions: (i) the fermion quantization of the ghosts will produce extra minus in front of the loop; (ii) the ghost field is complex, so there are two real degrees of freedom, instead of four in (11.24), hence

$$\Delta S_{\mathrm{ghost}} = \frac{N}{48\pi^2}\log\left(\frac{M_0^2}{\mu^2}\right)\frac{1}{4}\int dx\, G^a_{\mu\nu}(x)\, G^{\mu\nu\, a}(x)$$

$$\text{for ghosts}\,. \tag{11.25}$$

Now we approach the most important part of the calculation, the spin part of the gauge field loop. The last term in (11.20) is an "additional" vertex which does not interfere with the charge interaction. Hence, we must iterate it twice. This is trivially done with the knowledge we have already acquired (see page 92). We obtain

$$S + \Delta S_{\mathrm{spin}} = -\frac{1}{4g_0^2}\int dx\,\left(G^a_{\mu\nu}\, G^{\mu\nu\, a}\right)_{\mathrm{ext}} - iN\mathcal{I}\int dx\, G^a_{\mu\nu}\, G^{\mu\nu\, a}$$

$$= -\frac{1}{4g_0^2}\int dx\,\left(G^a_{\mu\nu}\, G^{\mu\nu\, a}\right)_{\mathrm{ext}}$$

$$+\frac{N}{4\pi^2}\log\left(\frac{M_0^2}{\mu^2}\right)\frac{1}{4}\int dx\, G^a_{\mu\nu}\, G^{\mu\nu\, a} \tag{11.26}$$

Assembling (11.24)–(11.26) we get the gluon loop as follows:

$$S + \Delta S = -\frac{1}{4g_0^2}\int dx\, G^a_{\mu\nu}\, G^{\mu\nu\, a}$$

$$+\frac{N}{4\pi^2}\left(1 - \frac{1}{12}\right)\log\left(\frac{M_0^2}{\mu^2}\right)\frac{1}{4}\int dx\, G^a_{\mu\nu}\, G^{\mu\nu\, a} \tag{11.27}$$

where 1 stands for the spin interaction (with its sign opposite to other contributions) while $-1/12$ represents the charge interaction combined with ghosts. Note that the charge interaction is twice larger than that of the ghosts and has the opposite sign. This just illustrates cancellation of unphysical degrees of freedom.

The final answer including the scalar and spinor matter in the fundamental representation (n_f and n_s flavors, respectively) is

$$S + \Delta S = -\frac{1}{4g_0^2}\int dx\, G^a_{\mu\nu}\, G^{\mu\nu\, a} \tag{11.28}$$

$$+\frac{1}{16\pi^2}\left(\frac{11\,N}{3} - \frac{2}{3}n_f - \frac{1}{6}n_s\right)\log\left(\frac{M_0^2}{\mu^2}\right)\frac{1}{4}\int dx\, G^a_{\mu\nu}\, G^{\mu\nu\, a}\,,$$

or, alternatively,

$$\frac{1}{g^2} = \frac{1}{g_0^2} - \frac{1}{16\pi^2}\left(\frac{11\,N}{3} - \frac{2}{3}n_f - \frac{1}{6}n_s\right)\log\left(\frac{M_0^2}{\mu^2}\right), \tag{11.29}$$

where n_f and n_s stand for the numbers of spinor and scalar flavors, respectively, in the fundamental representation of $\mathrm{SU}(N)_{\text{gauge}}$.

Fig. 11.2 Julian Schwinger.

Appendix 11.1: On Julian Schwinger and the background field method in QCD

The background field calculation method can be traced back to Julian Schwinger (1918-1994). Julian Schwinger received the 1965 Nobel Prize in theoretical physics. A discussion of development of this method in non-Abelian gauge field theories can be found in V. A. Novikov, M. A. Shifman, A. I. Vainshtein and V. I. Zakharov, *Calculations in External Fields in Quantum Chromodynamics. Technical Review,* Fortsch. Phys. **32**, 585 (1984).

Chapter 12

Lecture 12. What if Vacuum is Not Unique

Quantization of the gauge fields in the Higgs regime. Abelian fields (QED) and non-Abelian gauge fields (Yang-Mills). R_ξ gauge. The Green's function of ghosts in Yang-Mills theory in the Higgs regime.

12.1 Degenerate vacua: Path integral

In this section, I will show that one should deal with path integrals with extreme care, its naive treatment may lead to gross mistakes.

Indeed let us return to Chapter 3 and consider the double-well potential (see Fig. 12.1),

$$V(X) = \lambda \left(X^2 - \eta^2\right)^2 , \tag{12.1}$$

where we will normalize the coupling λ as follows:

$$8\lambda\eta^2 = \omega^2 . \tag{12.2}$$

Classically, this theory has two ground states, at $X = \pm\eta$. However, as well known from quantum mechanics, if the time interval $T \to \infty$, the ground state will be unique, representing a symmetric combination of the wave functions peaked at $X = \pm\eta$. The antisymmetric combination will be higher in energy. From this well-known consideration one might conclude that

$$\int X(t)\, \mathcal{D}X(t) \exp\left(iS[X(t)]\right) = 0 , \tag{12.3}$$

meaning that in the ground state $\langle X \rangle = 0$, which seems to be further enhanced by the Z_2 symmetry argument: the action is Z_2 symmetric under $X \to -X$ while the pre-exponent is antisymmetric under the above transformation.

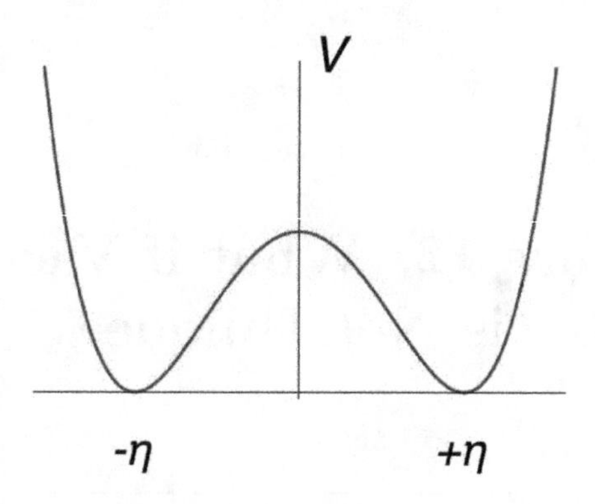

Fig. 12.1 Double-well potential.

Let us not make hasty conclusions, though. Assume that instead of quantum mechanics we now consider a field theory (say, in four dimensions) of a real scalar field $\phi(x)$ with the same potential,

$$V(\phi) = \lambda\left(\phi^2 - \eta^2\right)^2 . \tag{12.4}$$

If you apply the same Z_2 symmetry argument as above, you might say that at the quantum level, upon path-integration over all field configurations you arrive at $\langle\phi\rangle = 0$. Now, this will be a *wrong* answer!

If the spatial volume V is infinite – and that's what we usually assume in field theory – at the quantum level, we still end up with two distinct vacua, $\langle\phi\rangle \approx \pm\eta$. The discrete Z_2 symmetry is spontaneously broken, the vacua do not "mix" (more exactly, the probability of "mixing" is exponentially small in V and tends to zero as $V \to \infty$). Thus, a naive argument based on the Z_2 symmetry of the action which would lead, naively, to

$$\int \phi(x)\, \mathcal{D}\phi(x) \exp\left(iS[\phi(x)]\right) = 0\,, \tag{12.5}$$

is wrong. The correct path-integral strategy in this case is as follows. We should add a small Z_2 breaking term in the action, say $\int C\phi(x) d^4x$, then carefully calculate the path integral and observe that the result for (12.5) does *not tend to zero* in the limit $C \to 0$. Moreover, it must depend on the sign of C. For two different signs, we must arrive at two solutions corresponding to $\langle\phi\rangle \approx \pm\eta$.

A similar situation is valid in the case of spontaneous breaking of other global symmetries, not necessarily discrete. For continuous symmetries the vacuum manifold is continuously degenerate, and the order parameter will depend on "prehistory", i.e. on the way of introducing source terms with the subsequent elimination of those terms. Any point from the vacuum manifold may represent a valid vacuum.

12.2 Path-integral quantization in the Higgsed regime

If we consider a gauge theory with Higgsed gauge symmetries, then the vacuum is unique, generally speaking. However, in deriving the propagators for the massive gauge fields, ghosts and the Higgs fields we encounter a number of peculiarities which need to be discussed.

We know that in the Higgs regime there is a field rearrangement. Some components of the Higgs field are "eaten up" by the gauge field and converted into gauge field's longitudinal polarizations. This statement is valid in physical gauges, such as the unitary gauge.[1] However, the unitary gauge is inconvenient for quantum calculations. Say, the Feynman gauge would be much more practical. How can one properly derive the Feynman gauge? Below we will consider the gauge fixing procedure for arbitrary (covariant) gauges. In a generic gauge of this type, the would-be eaten components of the Higgs field remain "alive," and will be referred to as *unphysical Higgses.* One should distinguish them from the physical Higgs field. Unphysical Higgses appear only in loops and do not appear in the S matrix.

Our consideration will be conceptually similar to that of Chapter 9; technical details will be different, however.

As in Chapter 9, we will apply the Faddeev-Popov method. In this way we will define a class of gauges, called the R_ξ gauges. The unphysical Higgs fields are present in loops in the R_ξ gauges. These particles, along with the ghosts, will cancel the effects of the unphysical polarizations of the gauge field to maintain the unitarity of the theory.

12.2.1 *Abelian Higgsed theory*

The phenomenon of Higgsing was explained in Chapter 3. We will develop the R_ξ gauge first using the simplest example of the U(1) theory, with the Lagrangian (1.12),

$$\mathcal{L} = -\frac{1}{4} F_{\mu\nu}F^{\mu\nu} + (D_\mu\phi^\dagger)(D^\mu\phi) - V(|\phi|)\,, \tag{12.6}$$

[1]See page 114.

and

$$D_\mu = \partial_\mu - ieA_\mu \,. \tag{12.7}$$

Note that normalization of the A_μ field is changed compared to Chapter 2 to make its kinetic term canonic. Also normalization of v in (12.10) below is different compared to that in Chapter 2 by a factor $\sqrt{2}$ for the same reason.

Above, $\phi(x)$ is a complex scalar field. Let us represent it in terms of its real components,

$$\phi \equiv \frac{1}{\sqrt{2}}\left(\phi^1 + i\phi^2\right) . \tag{12.8}$$

A small gauge transformation can be written in the form [2]

$$\delta A_\mu = \partial_\mu \alpha(x) \,, \quad \delta\phi^1 = -e\alpha(x)\phi^2 \,, \quad \delta\phi^2 = e\alpha(x)\phi^1 . \tag{12.9}$$

Let us assume that $V(\phi)$ is chosen in such a way that the scalar field acquires a vacuum expectation value,

$$\langle \phi^1 \rangle = v \,. \tag{12.10}$$

Now we can separate the vacuum expectation value from the quantum fields as follows:

$$\phi^1 = v + h \,, \qquad \phi^2 = \varphi \tag{12.11}$$

where, as we will see later, h is the physical Higgs field while φ is an unphysical Higgs. We then rewrite the Lagrangian in terms of h and φ,

$$\mathcal{L} = -\frac{1}{4}\, F_{\mu\nu}F^{\mu\nu} + \frac{1}{2}(\partial_\mu h + eA_\mu\varphi)^2 + \frac{1}{2}\left(\partial_\mu\varphi - eA_\mu(v+h)\right)^2 - V(\phi) \,, \tag{12.12}$$

Note that the third term in (12.12) produces an "unwanted" mixing between $\partial_\mu\varphi$ and A^μ.

In order to define the functional integral over the variables h, φ, A_μ, we must introduce Faddeev-Popov gauge fixing. Starting from the functional integral

$$Z = N \int \mathcal{D}A\,\mathcal{D}h\,\mathcal{D}\varphi e^{iS[A,h,\varphi]} \tag{12.13}$$

we must introduce a gauge-fixing constraint as we did in Chapter 9, namely,

$$Z = N \int \mathcal{D}\alpha \int \mathcal{D}A\,\mathcal{D}h\,\mathcal{D}\varphi e^{iS[A,h,\varphi]}\delta^\infty(G^\alpha(A,h,\varphi)) \det \left|\left|\frac{\delta G^\alpha}{\delta\alpha}\right|\right| , \tag{12.14}$$

[2]The factor e in the transformation expressions for $\delta\phi^{1,2}$ are due to e in Eq. (12.7).

where N is a normalization constant and $(G(A,h,\varphi))$ is a gauge-fixing condition. Alternatively, we can introduce the gauge-fixing constraint as $\delta^\infty(G-\omega(x))$, see Eqs. (8.24) and (8.25), and integrate over $\omega(x)$ with a Gaussian weight, as in Chapter 8. Then we arrive at

$$Z = N\int \mathcal{D}\alpha \int \mathcal{D}A\,\mathcal{D}h\,\mathcal{D}\varphi \exp\left[i\int d^4x\left(\mathcal{L}[A,h,\varphi]-\frac{1}{2\xi}G^2\right)\right] \times \det\left|\left|\frac{\delta G^\alpha}{\delta\alpha}\right|\right| . \tag{12.15}$$

Now, to cancel the mixing term mentioned after (12.12) it will be wise to replace the gauge condition (8.23) by

$$G = \partial_\mu A^\mu + e\,\xi v\varphi\,, \tag{12.16}$$

where ξ is the same parameter as in (8.25) (Eq. (8.25) is kept intact).

Next, we observe with satisfaction that the quadratic terms of the gauged-fixed Lagrangian take the form

$$\mathcal{L}_{\rm quad} = -\frac{1}{2}A_\mu\left[-g^{\mu\nu}\partial^2 + \left(1-\frac{1}{\xi}\right)\partial^\mu\partial^\nu - e^2v^2g^{\mu\nu}\right]A_\nu + \frac{1}{2}(\partial^\mu h)^2 - \frac{1}{2}(m_h^2 h)^2 + \frac{1}{2}(\partial^\mu\varphi)^2 - \frac{\xi e^2}{2}v^2\varphi^2\,. \tag{12.17}$$

Here

$$m_h^2 = \frac{1}{2}V''\Big|_{\phi=(v+h)/\sqrt{2}}\,,\qquad m_\varphi^2 = \xi e^2v^2\,,\qquad m_A^2 = e^2v^2\,, \tag{12.18}$$

where the derivatives are taken with respect to h at $h=0$.[3]

The fact that m_φ^2 depends on the unphysical arbitrary parameter ξ shows that this Higgs is unphysical. The parameter ξ must disappear from any physical on-mass shell quantity. The φ field is never produced on mass shell.

To complete the Faddeev-Popov quantization procedure, we must derive the Lagrangian of the ghosts. This is done according to the standard rule,[4]

$$\frac{\delta G^\alpha}{\delta\alpha} \propto \partial^2 + \xi e^2 v(v+h)\,. \tag{12.19}$$

[3] m_φ^2 appears as the coefficient in front of $\varphi^2/2$ in the last term in (12.17) while m_A^2 appears in the last term of the first line in this expression.

[4] With our current normalization, the small gauge transformations take the form $\delta A_\mu = \partial_\mu\alpha$, and $\delta\phi = e\alpha(v+h)$.

The determinant of the above operator can be written as a ghost term in the Lagrangian,

$$\mathcal{L}_{\text{ghost}} = \bar{c}\left[-\partial^2 - \xi m_A^2\left(1+\frac{h}{v}\right)\right] c\,. \tag{12.20}$$

Even though the ghosts are sterile with respect to the gauge field (as expected in the Abelian theory), they interact with the physical Higgs field.

From the quadratic terms in the Lagrangian (see Eqs. (12.17) and (12.20)) we can readily find the propagators for these fields by inverting the appropriate quadratic differential operators (in the momentum space). The only relatively complicated case is that of the gauge field. Namely,

$$\begin{aligned}\langle A_\mu(k)A_\nu(-k)\rangle &= \frac{-i}{k^2-m_A^2}\left(g^{\mu\nu}-\frac{k^\mu k^\nu}{k^2}\right)+\frac{-i\xi}{k^2-\xi m_A^2}\left(\frac{k^\mu k^\nu}{k^2}\right)\\ &= \frac{-i}{k^2-m_A^2}\left(g^{\mu\nu}-\frac{k^\mu k^\nu}{k^2-\xi m_A^2}(1-\xi)\right),\end{aligned} \tag{12.21}$$

see also Problem 16 on page 263.

The Feynman gauge obviously corresponds to

$$\xi = 1\,. \tag{12.22}$$

Moreover, the propagator of the physical Higgs field is

$$\langle h(k)h(-k)\rangle = \frac{i}{k^2-m_h^2}\,, \tag{12.23}$$

while for the unphysical Higgs we get

$$\langle \varphi(k)\varphi(-k)\rangle = \frac{i}{k^2-\xi m_A^2}\,. \tag{12.24}$$

The ghost propagator takes the form

$$\langle c(k)\bar{c}(-k)\rangle = \frac{i}{k^2-\xi m_A^2}\,. \tag{12.25}$$

Note that at $\xi \to \infty$ the unphysical Higgs and the ghost disappear (since their mass terms tend to infinity). Simultaneously the gauge field propagator becomes remarkably simple,

$$\langle A_\mu(k)A_\nu(-k)\rangle = \frac{-i}{k^2-m_A^2}\left(g^{\mu\nu}-\frac{k^\mu k^\nu}{m_A^2}\right). \tag{12.26}$$

This is the *unitary gauge*. Unfortunately, the calculation of Feynman graphs with loops is practically impossible in the unitary gauge because of divergencies.

12.3 Non-Abelian case

Below we will consider the most practical case of the SU(2) gauge group spontaneously broken by a complex scalar field Φ^i in the fundamental representation, see Chapter 2. A more general consideration can be found in Peskin and Schroeder, pages 739–743. This is a part of the Glashow-Weinberg-Salam model. It would be also helpful to re-read Chapter 9. I remind that the vacuum expectation value of the field Φ^i can be always chosen in the following form:

$$\Phi_0 = \begin{pmatrix} v/\sqrt{2} \\ 0 \end{pmatrix}, \qquad v \text{ real and positive}. \tag{12.27}$$

Small deviation from the vacuum field above will be parametrized as

$$\Phi = \frac{1}{\sqrt{2}} \begin{pmatrix} v + h + i\chi^3 \\ -\chi^2 + i\chi^1 \end{pmatrix}, \tag{12.28}$$

where h is the physical Higgs field, while $\chi^{1,2,3}$ are unphysical Higgses which are eaten up by the vector bosons in the unitary gauge. From Chapter 2 we remember that all three gauge bosons are Higgsed and they acquire one and the same mass. In the expansion of the Lagrangian in the fields h and $\{\chi\}$ all three χ's do not enter the potential $V(\Phi^\dagger\Phi)$ at the quadratic level. Thus, all three χ's are classically massless.

Let us start with the kinetic term of the matter field *forgetting for a while about the physical Higgs h*, namely,

$$\Phi = \exp\left\{iT^a(2\chi^a/v)\right\}\Big|_{\text{linear}} \Phi_0\,. \tag{12.29}$$

The subscript in (12.29) indicates that the exponent must be expanded up to the linear order in χ's. All terms higher than linear should be discarded. Then we arrive at (12.28) with $h = 0$. In the case at hand the generator matrices $T^a = \frac{1}{2}\tau^a$ where τ^a are the Pauli matrices.

Keeping only the terms quadratic in the fields χ and A, it is not difficult to derive that

$$\begin{aligned}(D^\mu\Phi)^\dagger\,(D_\mu\Phi) \to\; & \frac{1}{2}\,(\partial_\mu\chi^a)^2 + g^2\Phi_0^\dagger T^a T^b \Phi_0\,(A_\mu^a\, A^{\mu\, b}) \\ & + ig\left(\Phi_0^\dagger T^a A_\mu^a\, \partial^\mu\Phi - \partial^\mu\Phi^\dagger T^a A_\mu^a \Phi_0\right) + \ldots \end{aligned} \tag{12.30}$$

Then the mass matrix of the Higgsed gauge bosons follows from the first line,

$$M_{ab}^2 = 2\,g^2\,\Phi_0^\dagger T^{\{a}T^{b\}}\Phi_0 = \frac{1}{4}g^2v^2\delta^{ab} \equiv m_A^2\delta^{ab}\,, \tag{12.31}$$

where $m_A = gv/2$. The braces in Eq. (12.31) mean symmetrization which is obviously necessary. Symmetrization automatically makes the mass matrix diagonal, with all equal eigenvalues. As we know from Chapter 2, this is the consequence of a hidden SU(2) "flavor" symmetry in the model at hand.

Now, let us deal with the second line in Eq. (12.30). Substituting Eqs. (12.27) and (12.28) (with $h = 0$) we arrive at

$$-g\,\frac{1}{2}\left(v\partial^\mu\chi^a\,A^a_\mu\right) \to g\,\frac{1}{2}\left(v\chi^a\,\partial^\mu A^a_\mu\right). \tag{12.32}$$

This is the analog of the cross term in Eq. (12.12). Now we can guess a gauge condition that will cancel the above cross term. Given that it appears in the Lagrangian in the form

$$\mathcal{L}_{\text{g. f.}} = -\frac{1}{2\xi}\left(G^a\right)^2, \tag{12.33}$$

we conclude that

$$G^a = \partial^\mu A^a_\mu + \frac{g}{2}\,\xi\,v\,\chi^a\,, \tag{12.34}$$

(cf. (12.16)) and

$$\mathcal{L}_{\text{g. f.}} = -\frac{1}{2\xi}\left(\partial^\mu A^a_\mu\right)^2 - \frac{1}{2}m_\chi^2\chi^a\chi^a - \frac{g}{2}\left(v\chi^a\,\partial^\mu A^a_\mu\right), \tag{12.35}$$

where

$$m_\chi^2 = \xi m_A^2\,. \tag{12.36}$$

I remind that the parameter m_A^2 was defined in (12.31).

The last step in our derivation is the determination of the ghost term. To this end we need to determine the response of (12.34) on the gauge transformation $U = \exp(iT^a\alpha^a)$ in the linear order in α^a. It is not difficult to establish that

$$\delta A^a_\mu = (D_\mu\alpha)^a\,,\quad \delta\chi^a = g\left(\frac{v}{2}\,\alpha^a + \frac{1}{2}\varepsilon^{acb}\chi^c\alpha^b\right), \tag{12.37}$$

implying that

$$\frac{\delta G^a}{\delta\alpha^b} \propto \left[-(\partial_\mu D^\mu)^{ab} - \frac{g^2}{4}\xi v^2\delta^{ab} - \frac{g^2 v\xi}{4}\varepsilon^{acb}\chi^c\right]. \tag{12.38}$$

As a result, the ghost term in the Lagrangian takes the form

$$\mathcal{L}_{\text{ghost}} = \bar{c}\left[-(\partial_\mu D^\mu)^{ab} - m_c^2\delta^{ab} - \frac{g^2 v\xi}{4}\varepsilon^{acb}\chi^c\right]c\,, \tag{12.39}$$

where

$$m_c^2 = \frac{g^2}{4}\xi v^2 = \xi\, m_A^2\,. \tag{12.40}$$

The quadratic part of the SU(2) Lagrangian with the Higgs mechanism is similar to that in Eq. (12.17), and so are the Green's functions (with the obvious extra color δ^{ab} factor). It is worth presenting the full Lagrangian term describing the unphysical Higgses,

$$\mathcal{L}_{\text{u. H}} = \frac{1}{2}\left[(\partial_\mu\chi)(D^\mu\chi) - \xi m_A^2\chi^2 + \frac{g^2}{4}(A_\mu)^2\chi^2\right]. \tag{12.41}$$

The above formalism as well as the R_ξ gauge were first worked out by G. 't Hooft, see Appendix 20.1 on page 215, whose contribution towards the birth of the Standard Model was absolutely instrumental.

Chapter 13

Lecture 13. Condensed Matter Week

Two of the most popular models in condensed matter physics are discussed at the introductory level. For thorough discussions see E. Fradkin, Field Theories of Condensed Matter Physics, Cambridge University Press, 2013.

13.1 General

In field theory we deal with one time dimension and d spatial dimensions, so that $D = 1 + d$. The path integral formulation is best suited for the Euclidean space, i.e. after the time rotation $t \to -i\tau$. In statistical physics time is irrelevant, so the functional integral refers to d dimensions. In this lecture my presentation of the Ising model will be based on the book by A. Polyakov [1], Sec. 3.1.

13.2 The (classical) Heisenberg model

The Heisenberg model [2] was introduced in 1928. Currently, the original model and its numerous modifications are used to describe a wide range of magnetic phenomena. Moreover, the continuous version of the model in two dimensions is used as an excellent theoretical laboratory. In the simplest case, the Hamiltonian can be written as

$$\mathcal{H} = -\sum \mathcal{J}_{ab}\, \vec{S}_a \vec{S}_b \tag{13.1}$$

with

$$\mathcal{J}_{ab} = \begin{cases} J \text{ if } a, b, \text{ are neighbors,} \\ 0 \text{ else.} \end{cases} \tag{13.2}$$

Here we consider a lattice,[1] and a set of three-component spin vectors $\vec{S} = \{\vec{S}_1, \vec{S}_2, \vec{S}_3\}$ of unit length,

$$\vec{S}^2 = 1, \tag{13.3}$$

each one placed on a lattice node and marked by the indices $a, b, \ldots$. The sum in (13.1) runs over all nodes. The spin vector is treated as classical but the interaction between them is quantum mechanical, according to (13.1).

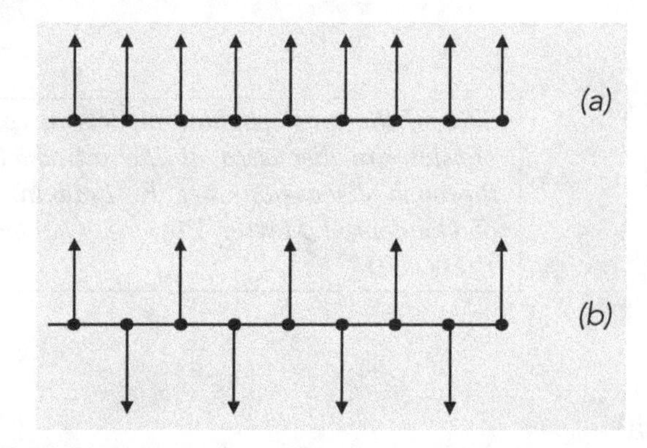

Fig. 13.1 The ground states of the Hamiltonian (13.1): (a) J positive, ferromagnetism; (b) J negative, anti-ferromagnetism.

If $J > 0$ the lowest-energy configuration is with all spins aligned (ferromagnetism, see Fig. 13.1(a)). In the opposite case $J < 0$, the lowest-energy configuration is with all spins anti-aligned (anti-ferromagnetism, see Fig. 13.1(b)).

The passage to the continuum limit is possible in the first case provided that the phenomena to be considered occur at distances much larger than the lattice site size ℓ. Then, for the nearest neighbors $\vec{S}_b = \vec{S}_a + \ell\partial\vec{S}_a + \frac{1}{2}\ell^2\partial^2\vec{S}_a +$ higher derivatives which I will ignore. Hence,

$$\begin{aligned} \vec{S}_a\vec{S}_b &= \vec{S}_a\vec{S}_a + \frac{1}{2}\partial\left(\vec{S}_a\vec{S}_a\right) + \frac{1}{2}\ell^2\vec{S}_a\partial^2\vec{S}_a + \ldots \\ &\to 1 + \frac{1}{2}\ell^2\vec{S}_a\partial^2\vec{S}_a\,. \end{aligned} \tag{13.4}$$

Replacing $J\ell^2$ by a continuous coupling constant $\frac{1}{g^2}$, we can write the Hamiltonian or the Euclidean Lagrangian as follows:

$$\mathcal{L} = \frac{1}{2g^2}\left(\partial_\mu\vec{S}\right)\left(\partial_\mu\vec{S}\right). \tag{13.5}$$

[1]We will limit ourselves to two-dimensional lattice for simplicity.

Superficially (13.5) looks as a free field theory. Do not forget, however, the constraint (13.3)! From this constrain we see that only two components of the vector $\vec{S}(x)$ are independent fields, the third component can be eliminated. Interactions will appear after this elimination is carried out. (See Problem 20 on page 265.) The model (13.5) is also referred to as the O(3) sigma model, to emphasize its symmetry.

The end-point of the unit vector $\vec{S}$ sweeps the two-dimensional sphere. Mathematicians would say that the target space of the model at hand is S^2, the two-dimensional sphere. This is the same as to say that the coordinate space is mapped onto S^2. It is well-known that S^2 has a nice geometrical representation through a stereographic projection, see Fig. 13.2.

The two-dimensional sphere is a very special manifold, it is a representative of a class of spaces called the *Kähler* spaces. The Kähler manifolds admit the introduction of complex coordinates, much in the same way as one can parametrize a two-dimensional plane by a complex number z and its complex conjugate $\bar{z}$. Now I will show how a complex field $\phi(x)$ can be introduced on S^2. If the original field $\vec{S}(x)$ was constrained, see Eq. (13.3), the field $\phi(x)$ has two components,

$$\phi_1(x) \equiv \operatorname{Re}\phi(x)\,, \qquad \phi_2(x) \equiv \operatorname{Im}\phi(x)\,, \tag{13.6}$$

which are unconstrained. This is convenient for the development of perturbation theory.

Figure 13.2 displays the target space sphere (with the unit radius), on which $\vec{S}$ "lives", and a ϕ plane which touches the sphere at the north pole. This is a two-dimensional plane which admits the introduction of the complex coordinate ϕ in a standard manner: if ϕ_1 and ϕ_2 are the Cartesian coordinates, then $\phi = \phi_1 + i\phi_2$. A ray of light emitted from the South pole pierces the sphere and the plane at the points denoted by small crosses. We then map these points onto each other,

$$S^1 = \frac{2\phi_1}{1+\phi_1^2+\phi_2^2}\,, \quad S^2 = \frac{2\phi_2}{1+\phi_1^2+\phi_2^2}\,, \quad S^3 = \frac{1-\phi_1^2-\phi_2^2}{1+\phi_1^2+\phi_2^2}\,. \tag{13.7}$$

The inverse transformation has the form

$$\phi = \frac{S^1 + i\,S^2}{1+S^3}\,. \tag{13.8}$$

It is clear that this is one-to-one correspondence. The only point which deserves a comment is the south pole ($S^1 = S^2 = 0\,,\ S^3 = -1$); it

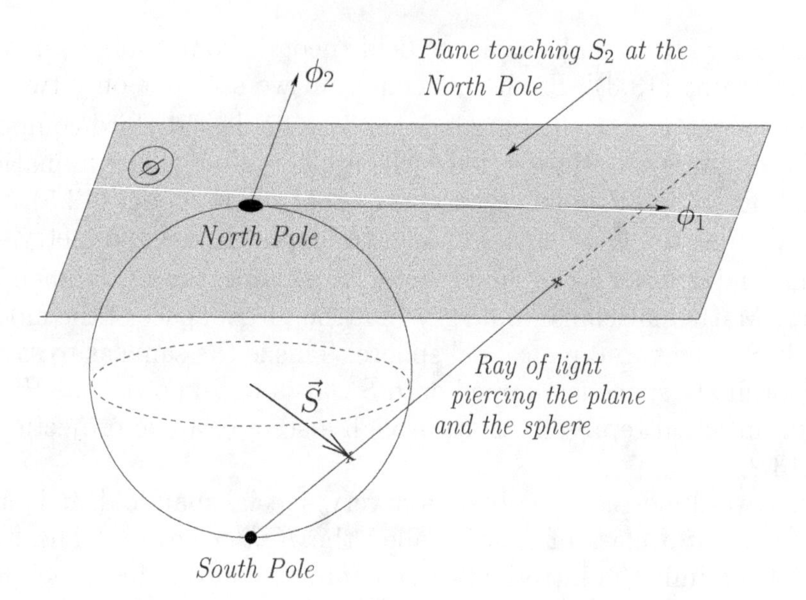

Fig. 13.2 Introduction of the complex coordinates on S^2 through the stereographic projection.

is mapped onto infinity. Since physically this is a single point on the target space, only such functions of ϕ are allowed for consideration that have a well-defined limit at $|\phi| \to \infty$, irrespective of the direction in the ϕ plane.

After a few quite simple but rather tedious algebraic transformations one obtains the action of the O(3) sigma model in terms of ϕ,

$$S = \int d^2x \, \frac{1}{(1+\bar{\phi}\phi)^2} \left\{ \frac{2}{g^2} \, \partial_\mu \bar{\phi} \, \partial^\mu \phi \right\}. \tag{13.9}$$

Let me note in passing that the expression $(1+\bar{\phi}\phi)^{-2}$ in front of the braces is nothing but the metric of the target space sphere in the given parametrization.

In this representation, the sigma model (13.9) is usually referred to as CP(1) model. CP stands for "complex projective."

Where have the symmetries of the original Lagrangian (13.5) gone? Only one global U(1) symmetry is apparent in Eq. (13.9),

$$\phi \to e^{i\alpha}\phi\,, \qquad \bar{\phi} \to e^{-i\alpha}\bar{\phi}\,. \tag{13.10}$$

Two other symmetries are realized nonlinearly,

$$\phi \to \phi + \epsilon + \bar{\epsilon}\,\phi^2\,, \qquad \bar{\phi} \to \bar{\phi} + \bar{\epsilon} + \epsilon\,\bar{\phi}^2\,, \tag{13.11}$$

where ϵ is a small complex parameter. To verify the invariance under (13.11), it is sufficient to observe that

$$\delta\,(1+\bar{\phi}\phi)^{-2} = -2\,(1+\bar{\phi}\phi)^{-2}\,(\epsilon\bar{\phi}+\bar{\epsilon}\phi)\,,$$
$$\delta\,(\partial_\mu\bar{\phi}\;\partial_\nu\phi) = (\partial_\mu\bar{\phi}\;\partial_\nu\phi)\,2\,(\epsilon\bar{\phi}+\bar{\epsilon}\phi)\,. \tag{13.12}$$

What is the vacuum (ground state) of the O(3) sigma model (13.5) or (13.9)? In perturbation theory, the O(3) symmetry of the Lagrangian is spontaneously broken, two massless excitations are present in the spectrum. However, beyond perturbation theory the answer crucially depends on the number of space-time dimensions. If $D \geq 3$, the O(3) symmetry remains spontaneously broken in the exact solution, SO(3)$\to$ SO(2) = U(1). Any point on S^2 can be chosen as a vacuum. S^2 then is referred to as the vacuum manifold. In two dimensions ($D = 2$) continuous symmetries cannot be spontaneously broken in interacting theories. This is the statement of the *Coleman* theorem. In this case the broken O(3) symmetry visible in perturbation theory is restored in the exact solution. The theory acquires a mass gap.

13.3 The Ising model *

This model presents a simplified "caricature" model of spin interactions. The elementary magnetic moments are assumed to be classical one-dimensional sticks which can point either upward (+) or downward (−). In other words, the spins are described by a discrete variable σ_i, which can take only two values, +1 and −1. This is clearly a gross idealization of spin. The variable σ_i lives on the lattice sites. In principle, one can consider a d-dimensional periodic lattice where d is arbitrary. In what follows we will limit ourselves to $d = 2$, i.e. two spatial dimensions.[2] This is called two-dimensional Ising model.

Let us assume that the size of the lattice link is unity. If the linear size of the lattice is L, the number of the lattice sites is $N = L^d$.

Any spin configuration is fully specified by a set of N numbers $\{\sigma_i\}$ where i runs over all lattice sites. Altogether there are 2^N spin configurations. The energy of the system in a given spin configuration is

[2] The three-dimensional case is clearly more relevant to nature, but we will not touch upon it here. For $d = 2$ the model is exactly solvable and is of immense pedagogical interest. The two-dimensional Ising model is arguably one of the most popular subjects for theoretical studies in statistical mechanics. Although it is 90-years old, a paper with the title containing "Ising Model" appears in ArXiv practically every week.

defined as

$$\mathcal{E}\{\sigma_i\} = -\sum_{i,j} J_{ij}\sigma_i\sigma_j - B\sum_i \sigma_i \tag{13.13}$$

Here B is an external (constant) magnetic field which may or may not be switched on. Usually one assumes that the interaction couples only the nearest neighbors, and that the interaction is isotropic, i.e. $J_{ij} \to J$ with $J > 0$. By measuring temperature in appropriate units we can always set the constant J equal to unity, as we will often do below.

The lattice appearing in the two-dimensional Ising model can be readily depicted on a sheet of paper (see Fig. 13.3). The partition function is

$$Z = \sum_{\{\sigma_i\}} \exp\left(-\beta\mathcal{E}\{\sigma_i\}\right) \tag{13.14}$$

where the sum in (13.14) runs over all possible spin configurations, and β is the inverse temperature, $\beta = 1/T$ (or, in the general case, $\beta = J/T$). Instead of the subscript i, I will label the spin variable by a vector $\vec{x}$ pointing to the corresponding lattice site,

$$\mathcal{E} = -\sum_{\vec{x},\vec{\delta}} \sigma_{\vec{x}}\,\sigma_{\vec{x}+\vec{\delta}}\,. \tag{13.15}$$

In two dimensions the number of the nearest neighbors is four; therefore the unit vector $\vec{\delta}$ takes only four different values: parallel or antiparallel to the x-axis and parallel or antiparallel to the y-axis. Two possible unit lattice vectors $\vec{\delta}$ are shown in Fig. 13.3. The magnetic field is set equal to zero in Eq. (13.15). In the absence of the magnetic field, our model possesses, at the microscopic level, a global Z_2 symmetry. Indeed, the sign of all spin variables can be simultaneously reversed,

$$\sigma_{\vec{x}} \to -\sigma_{\vec{x}}\,, \quad \forall\vec{x} \tag{13.16}$$

without changing the energy of any spin configuration.

Intuitively it seems quite clear that the model possesses two different phases. At low temperatures all spins are ordered – all point either upwards or downwards, ($\langle\sigma\rangle \neq 0$), while at high temperatures they are excited chaotically, and the system is disordered ($\langle\sigma\rangle = 0$). This intuitive expectation is confirmed by the exact solution of the model (see e.g. the book [3]). There exists a critical temperature at which a phase transition takes place.

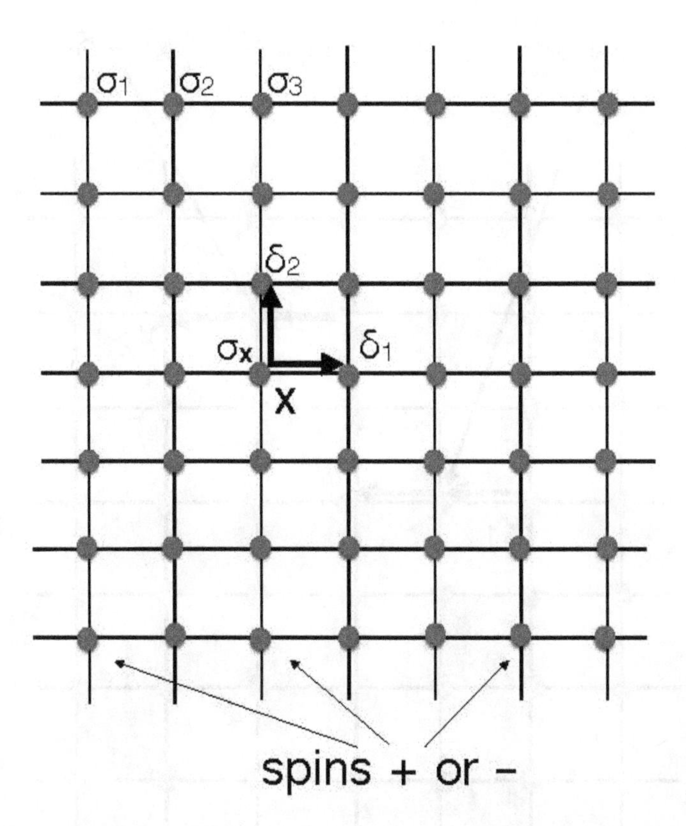

Fig. 13.3 The square lattice on which the two-dimensional Ising model is formulated. Each lattice site has unit length. The spin variables σ live on the nodes denoted by closed circles.

The discussion on the full exact solution is beyond the scope of the present lecture, but one can consider simple estimates illustrating the above assertion.

First of all, it is convenient to formulate the problem of the long range order (disorder) in the following terms. Let us assume that our system has a finite size and fix the value of $\sigma_{\vec{x}}$ at the boundary, say, by the condition

$$\sigma_{\vec{x}} = +1 \text{ if } \vec{x} \in \Gamma \tag{13.17}$$

where Γ is the boundary.

If the expectation value of the spin variable $\langle \sigma_{\vec{x}} \rangle$ taken at a point far from the boundary vanishes when the size of the system tends to infinity, there is no spontaneous magnetization. If, on the contrary, $\langle \sigma_{\vec{x}} \rangle$ stays finite in the limit of an infinitely large sample, the Z_2 symmetry of the model is spontaneously broken.

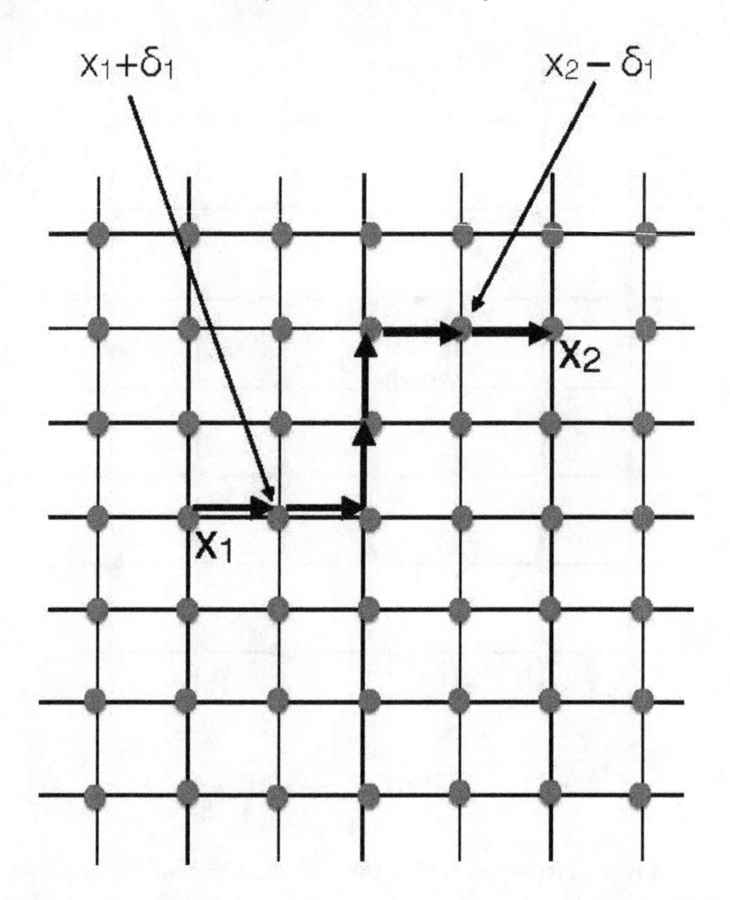

Fig. 13.4 The square lattice on which the two-dimensional Ising model is formulated.

Now, in the high temperature (small β) limit one can estimate the correlation function

$$\langle \sigma_{\vec{x}_1} \sigma_{\vec{x}_2} \rangle \tag{13.18}$$

by expanding the exponent $\exp(-\beta\mathcal{E})$ in Eq. (13.14). Due to the fact that we sum over all possible spin configurations (+1 and −1 at each lattice site) it is quite clear that the minimal number of terms to be kept is determined by the shortest path connecting the points $\vec{x}_1$ and $\vec{x}_2$,

$$\begin{aligned}\langle \sigma_{\vec{x}_1} \sigma_{\vec{x}_2} \rangle &= Z^{-1} \beta^R \sum_{\{\sigma_{\vec{x}}\}} \sigma_{\vec{x}_1} \sigma_{\vec{x}_1} \sigma_{\vec{x}_1+\vec{\delta}_1} \cdots \sigma_{\vec{x}_2-\vec{\delta}_n} \sigma_{\vec{x}_2}\, \sigma_{\vec{x}_2} \\ &= \beta^R \equiv \exp(-R/R_0)\,, \qquad R_0 = \frac{1}{\log(1/\beta)}\,, \end{aligned} \tag{13.19}$$

where R is the length of the shortest path (see Fig. 13.4).

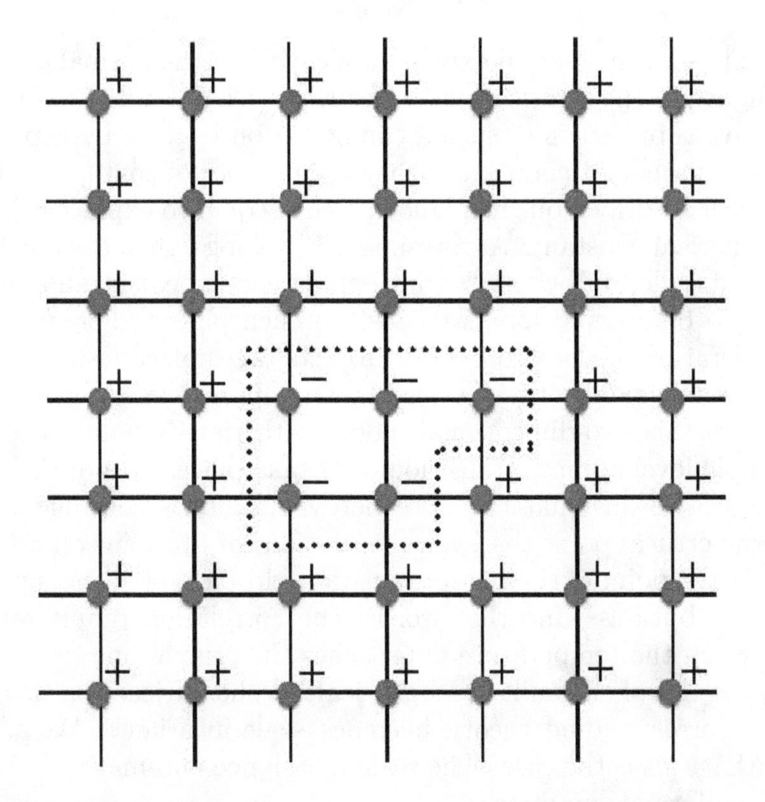

Fig. 13.5 A typical spin configuration at low temperatures. A droplet of "wrong spins" is shown by the dotted line.

Since the correlation length R_0 is small for small β the influence of the boundary condition (13.17) far away from the boundaries must also be small, and one can expect that

$$\langle \sigma_{\vec{x}} \rangle \sim \exp(-L/R_0) \to 0 \tag{13.20}$$

when the size of the system L tends to ∞. The qualitative argument presented above belongs to Polyakov [1].

Now, let us consider the inverse case of large β (small T) along the lines suggested by Rudolf Peierls [4].

A typical low temperature configuration is presented in Fig. 13.5. At small temperatures each "wrong" spin costs a lot in terms of $\beta\mathcal{E}$, and such configurations are heavily suppressed by $\exp(-\beta\mathcal{E})$. "Wrong" spins must be rare.

If a droplet of the "wrong" spins in the sea of the proper ones has the boundary of length ℓ, then, it is easy to check that for such a droplet we get $\exp(-\beta\ell)$ instead of $\exp(\beta\ell)$ appearing in the normal case of all spins aligned in the same direction, see Eqs. (13.14) and (13.15). Thus, the relative suppression of such a configuration is given by $\exp(-2\beta\ell)$. Moreover, mathematicians know that the number of contours of length ℓ which can be drawn on the plane lattice is equal to $\exp(\ell \log c)$ where c is a numerical constant. As a result, if $\beta > \frac{1}{2}\log c$ then the creation of the droplets filled in by the "wrong" spins is inexpedient and all spins turn out to be ordered (spontaneously broken phase). For $\beta < \frac{1}{2}\log c$ the proliferation of the wrong-spin droplets takes place resulting in the loss of the long range order (Z_2 symmetric phase).

Thus, in this two-dimensional model with the Z_2 symmetry at the microscopic level a phase transition does take place. It is of the *second order*, which means that the free energy per site is continuous in T, and at the critical point there are fluctuations of the spins on *all length scales*. At the point of the phase transition, droplets of all sizes produce equal contributions. In other words, the correlation radius tends to infinity when the temperature approaches the critical one.

The absence of the built-in length scale at the critical point suggests that the corresponding theory becomes scale-invariant. We will see shortly that this is the case. The scale invariance automatically implies the full conformal symmetry.

13.4 The Ising model at the critical point

Let us show that near the critical point, when the lattice formulation of the theory can be substituted by a continuous one, the two-dimensional Ising model is equivalent to the theory of a free Majorana fermion with mass $m \sim (T - T_c)$. The derivation presented below is due to A. Polyakov [1]. The idea is based on the transition from the original variables $\sigma_{\vec{x}}$ to the so-called disorder variables $\mu(\vec{y})$. The new variables are defined on the dual lattice (see Fig. 13.6); the points belonging to the dual lattice will be denoted by y_l (small crosses **x** in Fig. 13.6 are the nodes of the dual lattice).

In more concrete terms, the definition of $\mu(y)$ is as follows. Consider a point y_1 (point A in Fig. 13.6) and draw some path P from y_1 to infinity. Two possible paths, $ABC\infty$ and $ADC\infty$ are depicted in Fig. 13.6. Now, on all bonds which are intersected by the path P we change the sign of β and introduce a distorted partition function $\tilde{Z}(\vec{y}_1, P)$ analogously to Eq. (13.14), but with the negative β along the

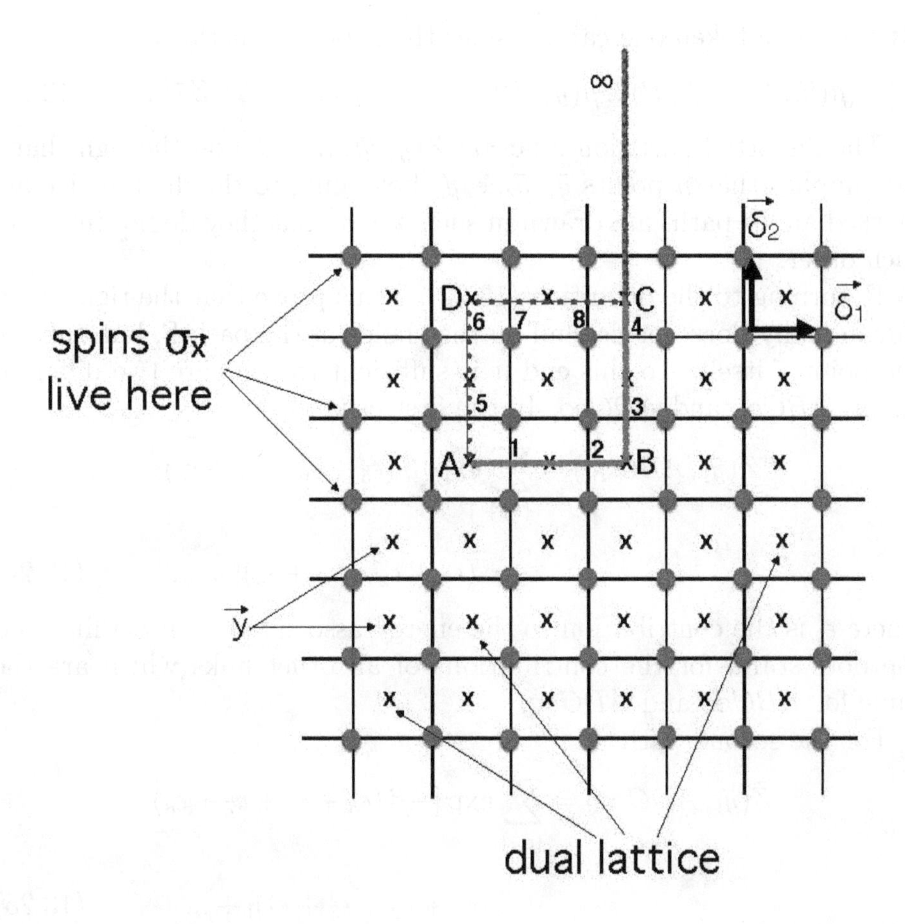

Fig. 13.6 The dual lattice on which the disorder parameter $\mu(y)$ is defined. The nodes of the dual lattice are depicted by small crosses **x**.

path P,

$$\tilde{Z} = \sum_{\{\sigma_i\}} \exp\left(\sum_{i,j,\,\text{n.n.}} \tilde{\beta}\, \sigma_i \sigma_j \right),$$

$$\tilde{\beta} = \begin{cases} -\beta \text{ for links intersected by } P, \\ \beta \text{ for links } \textit{not} \text{ intersected by } P, \end{cases} \tag{13.21}$$

see Fig. 13.6. Here n.n. means the nearest neighbors, as was discussed above.

The disorder variable $\mu(\vec{y}_1, P)$ is defined through its expectation value,

$$\langle \mu(\vec{y}_1, P) \rangle = \tilde{Z}(\vec{y}_1, P) Z^{-1} . \tag{13.22}$$

By the same token one can consider the n-point function

$$\langle \mu(\vec{y}_1, P)\, \mu(\vec{y}_2, P) ... \, \mu(\vec{y}_n, P) \rangle = \tilde{Z}(\vec{y}_1, \vec{y}_2 ... , \vec{y}_n, P) Z^{-1} \,. \qquad (13.23)$$

The distorted partition function $\tilde{Z}(\vec{y}_1, \vec{y}_2 ... \vec{y}_n, P)$ on the right-hand side implies that n points $\vec{y}_1, \vec{y}_2 ... , \vec{y}_n$ belonging to the dual lattice are marked and n paths are drawn in such a way that they do not intersect each other.

Returning to the definition (13.22), let us prove that the right-hand side actually does not depend on the choice of the path P, but only on the point $\vec{y}$ itself. To this end it is sufficient to compare two different paths, $ABC\infty$ and $ADC\infty$. In the first case

$$\tilde{Z}(\vec{y}_1, ABC\infty) = \sum_{\{\sigma_{\vec{x}}\}} \exp\left[\beta\left(\epsilon_5 + \epsilon_6 + \epsilon_7 + \epsilon_8\right)\right.$$

$$\left. - \beta\left(\epsilon_1 + \epsilon_2 + \epsilon_3 + \epsilon_4\right)\right] + ... \qquad (13.24)$$

where ϵ_i is the contribution to the energy associated with ith link and the dots stand for the contributions of all other links which are the same for $ABC\infty$ and $ADC\infty$.

For the second path

$$\tilde{Z}(\vec{y}_1, ADC\infty) = \sum_{\{\sigma_{\vec{x}}\}} \exp\left[-\beta\left(\epsilon_5 + \epsilon_6 + \epsilon_7 + \epsilon_8\right)\right.$$

$$\left. + \beta\left(\epsilon_1 + \epsilon_2 + \epsilon_3 + \epsilon_4\right)\right] + ... \qquad (13.25)$$

At first sight, the expressions in Eqs. (1.12) and (1.13) seem different, but one should not forget about the sum over all spin configurations. For each given configuration there is a "mirror" one when all spins inside the closed contour $ABCD$ are reversed, $\sigma_{\vec{x}} \to -\sigma_{\vec{x}}$ if $\vec{x} \in \square ABCD$. Under such a reversal $\epsilon_1,\ \epsilon_2, ... \,\epsilon_8 \to \big(-\epsilon_1,\ -\epsilon_2, ... -\epsilon_8\big)$, and we see that the modified partition functions in Eqs. (13.24) and (13.25) are identical.

The next step is to construct a field which has chances to behave like a free spinor field in the continuous limit. To this end, we fix a point $\vec{x}$ on the original lattice, and four surrounding points $\vec{y}_1, ...,\ \vec{y}_4$ of the dual lattice (Fig. 13.7). Then we define

$$\psi_a(\vec{x}) = \sigma_{\vec{x}} \mu(\vec{y}_a)\,, \quad a = 1, 2, 3, 4, \qquad (13.26)$$

assuming that the path connecting each point y_a with infinity goes in the horizontal direction to the left of $\vec{y}_a$.

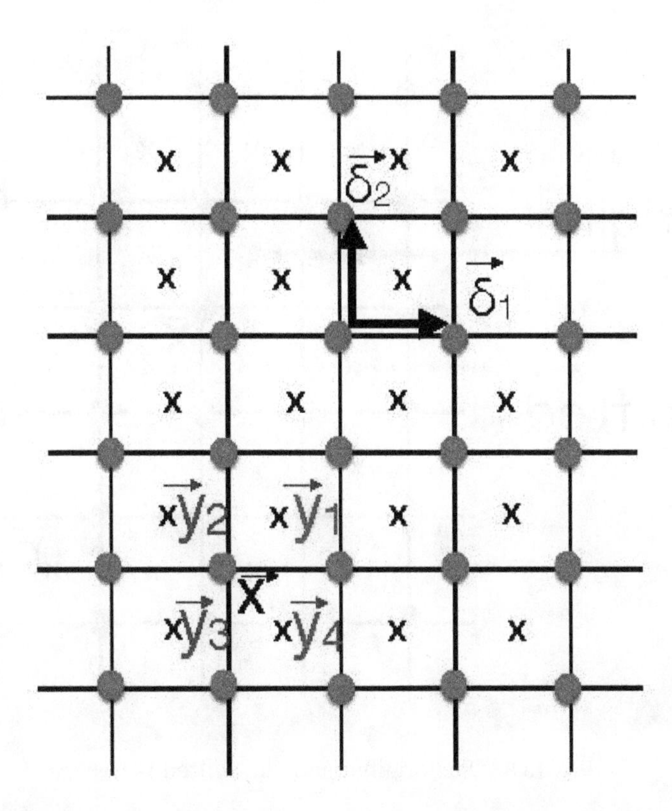

Fig. 13.7 Constructing the fermion field $\psi_a(\vec{x})$.

Now, it is easy to see that, say,

$$\psi_1(\vec{x}) = \sigma_{\vec{x}}\mu(\vec{y}_1) = \left\langle \sigma_{\vec{x}} \prod_{n=0}^{\infty} \exp\left(-2\beta\sigma_{\vec{x}-n\vec{\delta}_1}\sigma_{\vec{x}-n\vec{\delta}_1+\vec{\delta}_2}\right) \right\rangle \tag{13.27}$$

where the product reflects the fact that $\beta \to -\beta$ along the intersected bonds (see the definition of the disorder parameter μ above). Isolating the first term in the product and using the fact that $\sigma_{\vec{x}}^2 = 1$ we, obviously, get

$$\begin{aligned}
\langle\psi_1(\vec{x})\rangle &= \left\langle \sigma_{\vec{x}} \left[\prod_{n=1}^{\infty} \exp\left(-2\beta\sigma_{\vec{x}-n\vec{\delta}_1}\sigma_{\vec{x}-n\vec{\delta}_1+\vec{\delta}_2}\right)\right] \exp\left(-2\beta\sigma_{\vec{x}}\sigma_{\vec{x}+\vec{\delta}_2}\right) \right\rangle \\
&= \left\langle \sigma_{\vec{x}} \left[\prod_{n=1}^{\infty} \exp\left(-2\beta\sigma_{\vec{x}-n\vec{\delta}_1}\sigma_{\vec{x}-n\vec{\delta}_1+\vec{\delta}_2}\right)\right] \left[\cosh 2\beta - (\text{sinsh} 2\beta)\sigma_{\vec{x}}\sigma_{\vec{x}+\vec{\delta}_2}\right] \right\rangle \\
&= (\cosh 2\beta)\langle\sigma_{\vec{x}}\mu(y_2)\rangle - (\text{sinsh}2\beta)\langle\sigma_{\vec{x}+\vec{\delta}_2}\mu(y_2)\rangle \\
&= (\cosh 2\beta)\langle\psi_2(\vec{x})\rangle - (\text{sinsh}2\beta)\langle\psi_3(\vec{x}+\vec{\delta}_2)\rangle\,.
\end{aligned} \tag{13.28}$$

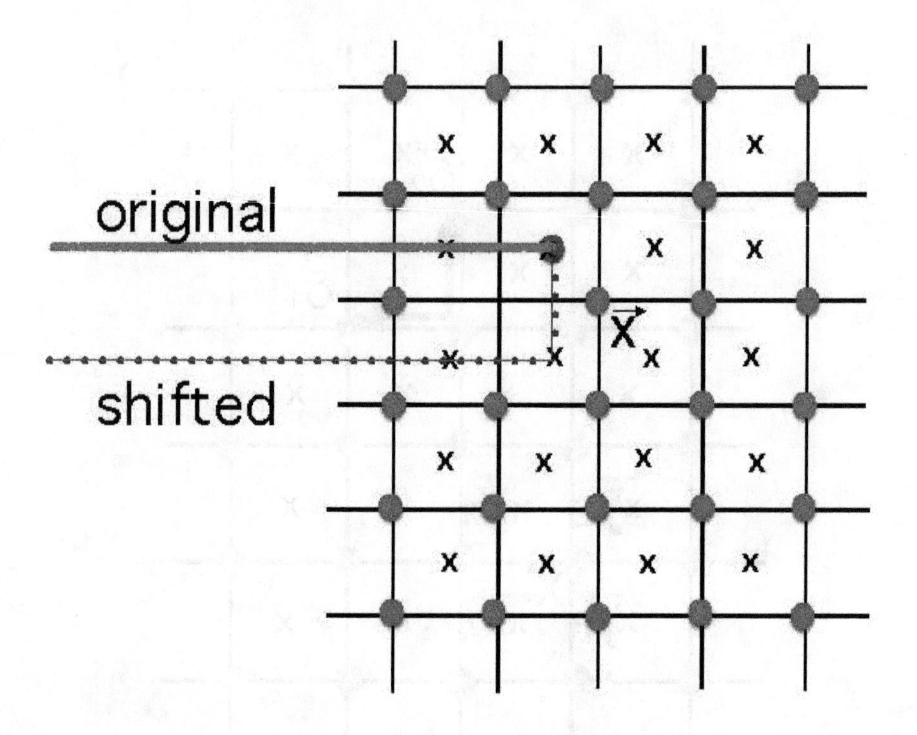

Fig. 13.8 The original and the shifted paths.

Likewise, we consider $\psi_2(\vec{x})$. Shifting the path as shown in Fig. 13.8 we then obtain

$$\langle\psi_2(\vec{x})\rangle = (\cosh 2\beta)\langle\psi_3(\vec{x})\rangle - (\sinh 2\beta)\langle\psi_4(\vec{x}+\vec{\delta}_3)\rangle\,, \qquad (13.29)$$

where $\vec{\delta}_3 = -\vec{\delta}_1$.

Furthermore,

$$\langle\psi_3(\vec{x})\rangle = (\cosh 2\beta)\langle\psi_4(\vec{x})\rangle + (\sinh 2\beta)\langle\psi_1(\vec{x}+\vec{\delta}_4)\rangle\,, \quad \vec{\delta}_4 = -\vec{\delta}_2\,, \qquad (13.30)$$

and

$$\langle\psi_4(\vec{x})\rangle = -(\cosh 2\beta)\langle\psi_1(\vec{x})\rangle + (\sinh 2\beta)\langle\psi_2(\vec{x}+\vec{\delta}_5)\rangle\,, \quad \vec{\delta}_5 = \vec{\delta}_1\,. \qquad (13.31)$$

The change of sign in the second term in Eq. (13.30) relatively to Eqs. (13.29), (13.27) occurs due to a trivial reason – instead of $\exp\left(-2\beta\sigma_{\vec{x}}\sigma_{\vec{x}+\vec{\delta}_2}\right)$ we have to expand now $\exp\left(2\beta\sigma_{\vec{x}}\sigma_{\vec{x}-\vec{\delta}_2}\right)$ because we add the link $-2\beta\sigma_{\vec{x}}\sigma_{\vec{x}-\vec{\delta}_2}$ in order to build up the needed tail and, correspondingly, subtract it.

The change of the sign in the first term of Eq. (13.31) is less trivial. Its origin as follows. We have shown above that distortions of the path (the "tail") do not affect the correlation functions of μ's. For ψ's the situation is the same if the distorted tail does not jump over the spin variable $\sigma_{\vec{x}}$ entering the definition of $\psi(\vec{x})$. It is easy to see (just repeating the argument given above) that if the distorted path jumps over $\sigma_{\vec{x}}$ there appears an extra minus sign. (Its origin is connected with the reversal of the sign of all spin variables inside the corresponding closed contour, see above.)

All four equations (13.28)–(13.31) can be nicely written in a unified form

$$\langle\psi_a(\vec{x})\rangle = (\cosh 2\beta)\langle\psi_{a+1}(\vec{x})\rangle - \sinh 2\beta\langle\psi_{a+2}(\vec{x}+\vec{\delta}_{a+1})\rangle \qquad (13.32)$$

provided that the following additional condition is imposed on $\psi_a(\vec{x})$:

$$\psi_{a+4}(\vec{x}) = -\psi_a(\vec{x})\,. \qquad (13.33)$$

This is our "master equation."

13.5 Solution

From the definition of the operator $\psi(\vec{x})$ (a spin variable $\sigma_{\vec{x}}$ plus a certain tail) we see that each transition $\psi_a \to \psi_{a+1}$ is just like a picture being rotated by $\pi/2$; passing from $\psi_1(\to \psi_2 \to \psi_3 \to \psi_4) \to \psi_5$ is equivalent to a total 2π rotation. The fact that this rotation introduces a sign,

$$\psi_5(\vec{x}) = -\psi_1(\vec{x})\,, \text{ etc.,} \qquad (13.34)$$

reminds us of fermions. Actually we will see that a fermion field does appear in the continuous limit.

Equation (13.33) can be used in order to determine the critical value of β. It is rather clear that at the point of the phase transition, where the correlation radius becomes infinite, there should exist an $\vec{x}$-independent solution. Taking into account the sign condition (13.33) we conclude that the solution, if present, should have the form

$$\psi_a \sim \exp(\pm i\pi a/4) \text{ or } \exp(\pm 3i\pi a/4)\,. \qquad (13.35)$$

Let us examine both the *ansätze* above one by one. The first *ansatz* is compatible with Eq. (13.32) provided that

$$\sinh 2\beta_c = 1, \quad \cosh 2\beta_c = \sqrt{2}\,. \qquad (13.36)$$

Indeed, substituting, say,

$$\psi_{a+2}(\vec{x}+\vec{\delta}_{a+1}) \to \psi_{a+2}(\vec{x})\,, \quad \psi_a \sim \exp(i\pi a/4)$$

in the master equation we reduce it to

$$e^{i\pi/4}\cosh 2\beta - e^{i\pi/2}\sinh 2\beta = 1\,,$$

which implies, in turn, Eq. (13.36). The second *ansatz* leads to no x-independent solution for real values of β. In other words, the mode proportional to $\exp(\pm i\pi a/4)$ becomes soft (long wavelength) near the critical point,

$$\beta = \beta_c = \frac{1}{2}\log(1+\sqrt{2})\,, \tag{13.37}$$

while the mode $\exp(\pm 3i\pi a/4)$ has characteristic wavelengths of the order of the lattice size and, thus, it does not survive in the continuum limit ($\beta \to \beta_c$) we are interested in.

From the heuristic consideration above it should be rather clear that near the critical point the continuum limit is recovered and the equation for the soft (long wave) mode is the Dirac equation. Specifically, let us write

$$\psi_a(\vec{x}) = u_+(\vec{x})\exp(i\pi a/4) + u_-(\vec{x})\exp(-i\pi a/4)\,,$$

$$u_-(\vec{x}) = \frac{1}{4}\sum_{a=1}^{4}\psi_a(\vec{x})\exp(i\pi a/4)\,,$$

$$u_+(\vec{x}) = \frac{1}{4}\sum_{a=1}^{4}\psi_a(\vec{x})\exp(-i\pi a/4)\,. \tag{13.38}$$

Convoluting both sides of the basic equation (13.32) with $\exp(\mp ia\pi/4)$ we project out $u_\pm(\vec{x})$ and obtain

$$(\partial_2 - i\partial_1)u_+ + 4e^{i\pi/4}(\beta-\beta_c)u_- = 0\,,$$

$$(\partial_2 + i\partial_1)u_- + 4e^{-i\pi/4}(\beta-\beta_c)u_+ = 0\,. \tag{13.39}$$

The phase of the mass term can be eliminated by the replacement

$$u_- \to u_- e^{-5\pi i/8}\,, \quad u_+ \to u_+ e^{5\pi i/8}\,. \tag{13.40}$$

Then Eqs. (13.39) take the form

$$(\partial_2 - i\partial_1)u_+ - mu_- = 0\,,$$

$$(\partial_2 + i\partial_1)u_- - mu_+ = 0\,. \tag{13.41}$$

where

$$m = 4(\beta - \beta_c) \tag{13.42}$$

Equations (13.41) can be obtained from the Lagrangian (I use field-theoretic language)

$$\mathcal{L} = \bar{U}\left(\sum_{j=1,2} -i\,\partial_j\gamma_j + im\right)U\,, \tag{13.43}$$

where

$$\gamma_1 = \sigma_2\,, \quad \gamma_2 = \sigma_1\,,$$

σ_j are the Pauli matrices and

$$U = \begin{pmatrix} u_- \\ u_+ \end{pmatrix}, \quad \bar{U} = (u_+\,,\ -u_-)\,. \tag{13.44}$$

The Dirac spinor Ψ in two dimensions has two independent components; moreover, $\bar{\Psi}$ is an independent integration variable with two components too. Altogether we have four components. For the problem on hand, we have only two independent components, see Eq. (13.44). Thus, we deal with the Majorana fermion.

Thus far we have considered only the single field average $\psi_a(\vec{x})$ and obtained the effective Lagrangian (13.43) on the basis of this analysis. One may worry whether this Lagrangian exhausts all physics in the sense that nothing new emerges when considering more complicated correlation functions, say,

$$\langle\psi_{a_1}(\vec{x}_1), ..., \psi_{a_n}(\vec{x}_n)\rangle\,. \tag{13.45}$$

Actually, it is not difficult to show that with respect to each argument $\vec{x}_1, ..., \vec{x}_n$, these correlation functions satisfy the same equation (13.32), with only one evident modification: on the right-hand side, some contact terms appear. In the continuum limit the latter reduce to usual delta-functions, just the same δ-function that one obtains in the theory with the Lagrangian (13.43).

Now we know enough to do an instructive exercise – the calculation of the specific heat,

$$C = T\frac{\partial^2 F}{\partial T^2} \tag{13.46}$$

where F denotes free energy. In the field-theoretic language we have to calculate the derivative of the vacuum energy density E, namely

$$C \sim \left.\frac{\partial^2 E}{\partial T^2}\right|_{T\to T_c} \tag{13.47}$$

where

$$E \sim -V^{-1} \log Z \tag{13.48}$$

It is obvious that at T close to T_c

$$dT \sim \beta^{-2} d\beta = \beta_c^{-2} d\beta \sim d\beta \tag{13.49}$$

and

$$dm \sim d\beta\,, \tag{13.50}$$

see Eq. (13.42). Hence,

$$C \sim \left.\frac{\partial^2 E}{\partial m^2}\right|_{T\to T_c} . \tag{13.51}$$

As we will see in one of the subsequent lectures, in field-theoretic language

$$E \sim m\langle \bar{U} U\rangle\,. \tag{13.52}$$

The experience we have obtained from field theory prompts us the way to move further in computing (13.52). At $T \to T_c$, as long as one is interested only in the most singular term,

$$\langle \bar{U} U\rangle \sim \int \frac{d^2 p}{(2\pi^2)}\, i\,\mathrm{Tr}\left[\frac{\not{p} + im}{p^2+m^2}\right] \sim m \log\left(\frac{M_0}{m}\right)$$

Then, taking into account the relation (13.51) we get

$$C \sim \frac{\partial^2}{\partial m^2}\left[m^2 \log\left(\frac{M_0}{m}\right)\right]$$

$$\sim \log\left(\frac{1}{m}\right) \sim \log\left|\frac{\beta-\beta_c}{\beta_c}\right| . \tag{13.53}$$

Thus, we reproduce, in a rather trivial way, the famous Onsager's result: the logarithmic dependence of specific heat on $(T - T_c)$. A field theorist would express the very same computation of the very same quantity by the following words: the vacuum energy density depends on the mass parameter as $m^2 \log m$. We see that the field theory practitioners and those who are in statistical mechanics do the same things but use different languages for the description of their results. It is quite clear that those who know both languages can benefit a lot by applying the knowledge obtained in one field to the other.

The most important lesson for us here is that the two-dimensional Ising model at the critical point $T = T_c$ reduces to the continuous theory of a free *massless* Majorana fermion.[3]

The two-dimensional Ising model at criticality is, probably, the most well-known example of how the description of the critical phenomena in solid state physics gives rise to conformal field theories.

Appendix 13.1: On Ernest Ising

Since the two-dimensional Ising model became one of the landmarks in theoretical physics, a few historical remarks are in order here. In fact, the concept of the model was first suggested by Wilhelm Lenz [5], the thesis supervisor of E. Ising (the same Lenz who introduced the "Runge-Lenz" vector in the context of a hidden symmetry in the quantum-mechanical problem of the hydrogen atom). Lenz asked his student, Ernest Ising, to study the model in connection with a possible description of ferromagnetism; Lenz's role in the further development of the model was marginal.

Ising solved the model in the case $d = 1$ [6], and showed that in one dimension the phase transition from para- to ferromagnetism does not take place. Although in one dimension this is certainly true, Ising somehow talked himself into believing that the same situation, no phase transition, must persist in two or three dimensions, which was wrong. Ising got so frustrated because of the "uselessness" of his model that he left theoretical physics and became an educator. He barely escaped from the Nazis in Germany in 1939. He survived during the WWII in Luxemburg thanks to his wife Johanna (see her recollections, Jane Ehmer Ising, *Walk on a Tightrope*). With no contacts with the theoretical physics community Ising did not know about the revolutionary developments around his model. He learned about them only in 1947 when he came to America where he got a teaching position.

In the mid-1930s the model was studied [7] in the mean field approximation, which was later improved by Hans Bethe [8]. The first exact result was obtained in the two-dimensional Ising model by Hans Kramers and Gregory Wannier [9] who made heavy use of the so-called Kramers-Wannier duality: a symmetry between the high- and low-T expansions of the partition function. This symmetry allowed them to

[3] This is a classical example of the conformally invariant theory corresponding to the value of the so-called Virasoro central charge $c = \frac{1}{2}$.

Fig. 13.9 Ernest Ising.

calculate the phase transition temperature. The culmination of the story is certainly Lars Onsager's exact solution of the model (with the magnetic field switched off) which was announced by Onsager at a meeting of the New York Academy of Sciences in 1942 and published in 1944 [10]. Further developments are due to Kaufman, Onsager and Yang (see [11]). In particular, the exact spin correlation functions were found in 1949 [12].

A. B. Pippard wrote in 1961 in a private communication to the Royal Society,

> Onsager's exact treatment, which created a sensation when it appeared, showed that the specific heat in fact rose to infinity at the transition point, a phenomenon which profoundly disturbed those who were sure that fluctuations always smoothed over the asperities which were created by approximations in the analysis. This work gave a new impetus to the study of cooperative phenomena,... and it is certainly the most important single achievement in this important field.

References

[1] A. Polyakov, *Gauge Fields and Strings*, (Harwood Academic Publishes, Chur, 1987).

[2] Werner Heisenberg, "Zur Theorie des Ferromagnetismus," Zeitschrift für Physik **49**, 619 (1928) [https://doi.org/10.1007/BF01328601].

[3] Rodney J. Baxter, *Exactly Solved Models in Statistical Mechanics*, (Dover Publications, 2008).

[4] R. Peierls, Proc. Camb. Phil. Soc. **32**, 471; 476 (1936).

[5] W. Lenz, Z. Phys. **21**, 613 (1920).

[6] E. Ising, Z. Phys. **31**, 253 (1925).

[7] W. Bragg and E. Williams, Proc. Roy. Soc. **A145**, 199 (1934); **A151**, 540 (1935).

[8] H. A. Bethe, Proc. Roy. Soc. **A150**, 552 (1935).

[9] H. A. Krammers and G. H. Wannier, Phys. Rev. **60**, 252; 263 (1941).

[10] L. Onsager, Phys. Rev. **65**, 117 (1944).

[11] B. Kaufman, Phys. Rev. **76**, 1232 (1949); L. Onsager, Nuov. Cim. Suppl. **6**, 261 (1949); C. N. Yang, Phys. Rev. **85**, 808 (1952).

[12] B. Kaufman and L. Onsager, Phys. Rev. **76**, 124 (1949).

Chapter 14

Lecture 14. Renormalization Group; RG Flow

Magnetic moment vs. charge interactions. The Gell-Mann–Low functions (currently known as β functions). Asymptotic freedom versus infrared freedom. Zeros of the β functions. Asymptotic flow to the zeros of the β functions. Conformal symmetry. A qualitative explanation of asymptotic freedom.

14.1 Charge renormalization (*continued*)

In Chapter 13 the expression for the charge renormalization of Yang-Mills field interacting with itself and matter was derived,

$$\frac{1}{g^2} = \frac{1}{g_0^2} + \frac{1}{16\pi^2}\left[\left(-4+\frac{1}{3}\right)N + \frac{2}{3}n_f + \frac{1}{6}n_s\right]\log\frac{M_0^2}{\mu^2}. \tag{14.1}$$

In the first bracket (-4) came from the spin-dependent interaction with magnetic moment of the vector (i.e. spin 1) particle, while $+\frac{1}{3}$ from the charge interaction.

Actually, we can separate spin and charge interactions in the fermion contribution as well. It corresponds to rewriting of $2/3$ as

$$\frac{2}{3} = 1 - \frac{1}{3}. \tag{14.2}$$

Indeed, $-\frac{1}{3}$ refers to the charge part: it is twice the scalar by magnitude because of two polarization states, and its opposite sign reflects minus for the fermion loop. The spin part is $(+1)$.

Thus we see that the ratio of spin part to the charge part is described by

$$\frac{\text{spin part}}{\text{charge part}} = -12s^2 \tag{14.3}$$

for both, spinor, $s = 1/2$, and vector, $s = 1$, cases. It naturally reflects the proportionality of magnetic moment to spin s. The magnetic interaction enters quadratically in the one-loop calculation.

Note, that the number of polarization states is two for both, the Dirac spinor and the massless vector. Note also an integer nature of $(+1)$ and (-4) in their spin parts. This is not accidental, it reflects a geometrical nature of those numbers.

As for dependence on representation, it is convenient to rewrite Eq. (14.1) as

$$\frac{1}{g^2} = \frac{1}{g_0^2} + \frac{1}{16\pi^2}\left[-\frac{11}{3}T_G + \frac{4}{3}T_R\, n_f + \frac{1}{3}T_R\, n_s\right]\log\frac{M_0^2}{\mu^2}, \tag{14.4}$$

where T_R is the Dynkin index of the representation R defined as

$$\mathrm{Tr}\, T_R^a T_R^b = T_R\,\delta_{ab}\,. \tag{14.5}$$

Here T_R^a are generators in the R representation. For the SU(N) fundamental representation $T_{\text{fund}} = 1/2$, and for the adjoint, denoted by G, the index $T_G = N$.

14.2 Running couplings, renormalization group

The most general approach to renormalization was formulated by Wilson [1]. The starting point is an introduction of the local Lagrangian density $\mathcal{L}$ which can be written as

$$\mathcal{L}(x) = \sum c_i\, \mathcal{O}_i(x)\,, \tag{14.6}$$

where the operators $\mathcal{O}_i(x)$ are expressed via local fields and their derivatives while the coefficients c_i can be viewed as generalized couplings. Actually, the existence of the ultraviolet divergences implies that we have to introduce also the UV limit M_0 in the Euclidean momentum space.[1] Physically, it means that we do not consider distances shorter than $1/M_0$.

Thus, we can say that the initial Lagrangian $\mathcal{L}$ is local only up to distances $1/M_0$. This can be marked as

$$\mathcal{L}|_{M_0}(x) = \sum c_i(M_0)\, \mathcal{O}_i|_{M_0}(x)\,. \tag{14.7}$$

[1]Sometimes I denote the UV cut-off as M_{UV}.

On the other hand, we would like to have physical results at low energies to be independent of a particular choice of the UV cutoff M_0. This means that we can go to a different value $\widetilde{M}$ of the cutoff,

$$\mathcal{L}|_{\widetilde{M}}(x) = \sum c_i(\widetilde{M})\, \mathcal{O}_i|_{\widetilde{M}}(x) \tag{14.8}$$

without changing the physical amplitudes.

In terms of the path integral representation transition from $\mathcal{L}|_{M_0}$ to $\mathcal{L}|_{\widetilde{M}}$ means that for $\widetilde{M} < M_0$, we integrate over the fields with Euclidean momenta p in the interval $\widetilde{M} \leqslant p \leqslant M_0$. Independence on M_0 and $\widetilde{M}$ is due to the fact that we are just reshuffling this range of momenta moving it from matrix elements of operators $\mathcal{O}_i$ into the coefficients c_i. This integration defines $c_i(\widetilde{M})$ via $c_i(M_0)$. Considering the limit of small separation between $\widetilde{M}$ and M_0, we can say that the variation $M_0 dc_i(M_0)/dM_0$ is a certain function of all coefficients $c_i(M_0)$. Equivalently, denoting the sliding scale $\widetilde{M}$ by μ, we conclude that

$$\mu\,\frac{dc_i(\mu)}{d\mu} = \beta_i\big(\{c_j(\mu)\}\big)\,. \tag{14.9}$$

Here $(\{c_j(\mu)\}$ in the argument of β_i denotes the whole set of all coefficients $c_j(\mu)$. The set $\{\beta_i\}$ sometimes is referred to as generalized β functions.

Equations (14.9) constitute equations of the renormalization group which express independence of physics on the sliding scale μ. They say that μ separates short distances, $\leqslant 1/\mu$, and large distances, $\geqslant 1/\mu$. The former go into the coefficients $(\{c_j(\mu)\}$ while the latter define the set of operators $\{\mathcal{O}_j(\mu)\}$. The scale of separation μ is called the *normalization point.*

14.3 Running gauge coupling. Beta function for Yang-Mills theory with matter

The relevant coupling in this case is

$$\alpha(\mu) = \frac{g^2(\mu)}{4\pi}\,, \tag{14.10}$$

and the renormalization group (RG) equation takes the form

$$\mu\,\frac{d\alpha}{d\mu} = \beta(\alpha)\,, \tag{14.11}$$

where the β function is also known as the Gell-Mann–Low function. Murray Gell-Mann and Francis Low invented this concept. In this section, we will discuss quantum chromodynamics as an example.

Previously we calculated the one-loop expression for $\alpha(\mu)$,

$$\alpha(\mu) = \frac{\alpha_0}{1 - b\,(\alpha_0/4\pi)\log M_0^2/\mu^2}\,, \tag{14.12}$$

where the parameter b is

$$b = \frac{11}{3}N - \frac{2}{3}n_f - \frac{1}{6}n_s\,. \tag{14.13}$$

It is easy to see that $\mu d\alpha/d\mu$ is the function of $\alpha(\mu)$ indeed,

$$\mu\,\frac{d\alpha}{d\mu} = -b\,\frac{\alpha^2}{2\pi}\,. \tag{14.14}$$

This shows the RG equation is satisfied and in the leading (one-loop) order the β function reduces to

$$\beta(\alpha) = -b\,\frac{\alpha^2}{2\pi}\,. \tag{14.15}$$

The RG equation proves that the expression (14.12) works beyond the second order in α_0: it sums up terms $\alpha_0(\alpha_0\log M_0/\mu)^n$ to all orders. Two-loop calculations of $\alpha_0^3\log M_0/\mu$ will determine α^3 terms in the β function, and so on. Also, in the case of scalars, two-loop terms $\alpha^2\lambda$ will show up in β. Here the coupling λ describes self-interaction of scalars, $\lambda(\phi^\dagger\phi)^2$. Of course, it comes together with the appropriate RG equation for $\lambda(\mu)$ itself.

14.3.1 *Asymptotic freedom in Yang-Mills theories*

A special (and rather unique) feature of Yang-Mills theory is the negative sign of the β function provided that the number of matter fields is not to large. The negative sign means that the effective coupling decreases at larger μ, i.e. at shorter distances.

Of course, it is also visible from the explicit form of $\alpha(\mu)$,

$$\alpha(\mu) = \frac{\alpha(\mu_0)}{1 + b\,(\alpha(\mu_0)/4\pi)\log \mu^2/\mu_0^2}\,. \tag{14.16}$$

where we fix α at the point μ_0. This expression can be also written as

$$\alpha(\mu) = \frac{2\pi}{b \log \mu/\Lambda}, \tag{14.17}$$

where

$$\Lambda = \mu_0 \exp\left[-\frac{2\pi}{b\alpha(\mu_0)}\right] \tag{14.18}$$

is the RG invariant. The emergence of the dynamical scale Λ from the dimensionless coupling constant (and a hidden parameter, the UV cut-off, which is not seen in the Lagrangian) is called *dimensional transmutation.*

The phenomenon of fall-off of the running coupling at short distances goes under the name of *asymptotic freedom* (AF) [2]. AF of quantum chromodynamics, the theory of strong interaction, allows one to carry out perturbative calculations at short distances, or, "deep virtualities" (i.e. $\mu \gg \Lambda$). The interaction becomes strong at $\mu \sim \Lambda$, see (14.17). At distances $\sim \Lambda^{-1}$ the perturbative regime fails, and nonperturbative physics (i.e. physics of quark confinement takes over, see Fig. 14.1).[2]

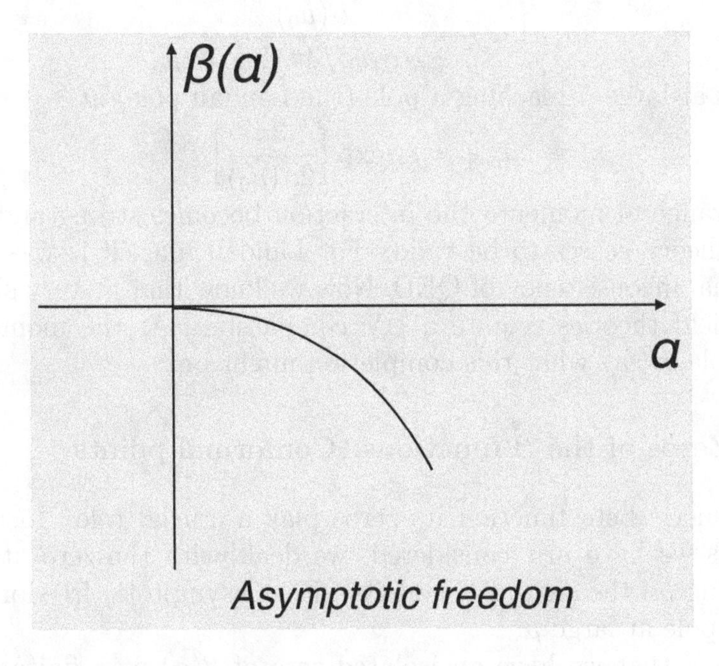

Fig. 14.1 The β function of an asymptotically free theory.

[2]The first analytic proof of quark confinement was found in [3].

14.4 QED, the Landau pole

What happens in case of positive β function? This occurs quite often, for instance, in spinor QED

$$-b = \frac{4}{3}\,. \tag{14.19}$$

The corresponding plot is presented in Fig. 14.2. Theories with the positive β functions are called infrared (IR) free because the running constant $\alpha(\mu)$ falls off at small μ (i.e. large distances) and grows at short distances. This pattern is opposite to that we saw in the asymptotically free field theories. Yang-Mills theory is, actually, quite a rare exception. Indeed, the negative sign of b means that the charge we see at large distances is smaller that the one at shorter distances, so we deal with a "screening" of the original charge (for more details see Appendix 14.1 on page 149). The screening due to vacuum polarization seems quite generic. Indeed, the component of the virtual pair with the charge opposite to the probe would go to the probe due to attraction while the other component of the pair would go to large distances.

The expression for $\alpha = e^2/4\pi$ in the spinor QED is $(\mu, \mu_0 \gg m_e)$

$$\alpha(\mu) = \frac{\alpha(\mu_0)}{1 - (\alpha(\mu_0)/3\pi)\log \mu^2/\mu_0^2}\,. \tag{14.20}$$

It grows at large μ reaching a pole (the Landau pole) at

$$\mu_{\text{pole}} = \mu_0 \exp\left[\frac{3\pi}{2\alpha(\mu_0)}\right]. \tag{14.21}$$

In this range of momenta the interaction becomes strong and perturbation theory ceases to be valid. For Landau himself it was a signal of internal inconsistency of QED. Now we know that at very short distances field theories require a UV completion. At the moment it is impossible to say what this completion might be.

14.5 Zeros of the β functions. Conformal points

For a generic beta function its zeros play a crucial role. In one-loop examples we have just considered, we dealt with the zero at $\alpha = 0$. Depending on the sign of b we have either asymptotic freedom or the Landau pole at large μ.

Imagine that we have an isolated zero of $\beta(\alpha)$ at a finite value of $\alpha = \alpha_*$. Expanding near α_*, we approximate β as

$$\beta(\alpha) = \eta\,(\alpha - \alpha_*)\,, \tag{14.22}$$

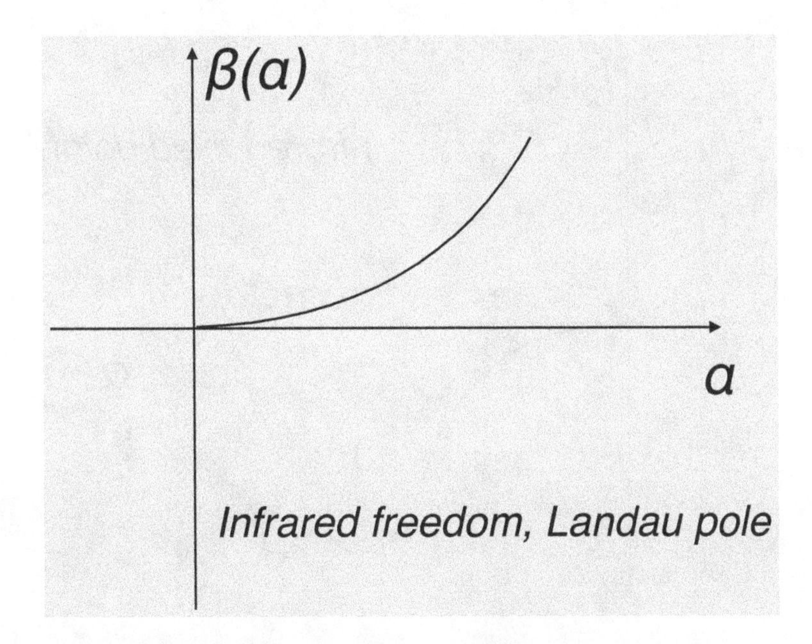

Fig. 14.2 The β function of an infrared free theory.

where η is some number (positive in the case at hand). Equation (14.22) assumes the zero is of the first order. It is simple to solve the RG equation (14.11) for such β. Let us start from $\alpha(\mu_0) < \alpha_*$. Correspondingly, we will consider $\mu < \mu_0$

$$\alpha_* - \alpha(\mu) = \left[\alpha_* - \alpha(\mu_0)\right] \left(\frac{\mu}{\mu_0}\right)^\eta . \tag{14.23}$$

As μ becomes $\ll \mu_0$, we see that $\alpha(\mu)$ approaches α_* from below. Choosing the initial condition $\alpha(\mu_0) > \alpha_*$ we will see that in the IR limit, $\alpha(\mu)$ will approach α_* from above. In both cases the degree of approach is power-like, i.e. fast, much faster that the logarithmic fall-off of $\alpha(\mu)$ in the UV limit.

In general, if η is positive, we see that at decreasing μ the coupling α approaches α_* in the power-like manner. This is a conformal behavior in the IR at the point α_*; this point is thus IR stable, see Fig. 14.3. However, it is unstable in the UV, i.e. when $\mu \gg \mu_0$. The situation is reversed for negative η. The fixed point in this case will be UV stable and IR unstable. This might happen in QED if a zero in the QED β function could be found at a non-vanishing value of α.

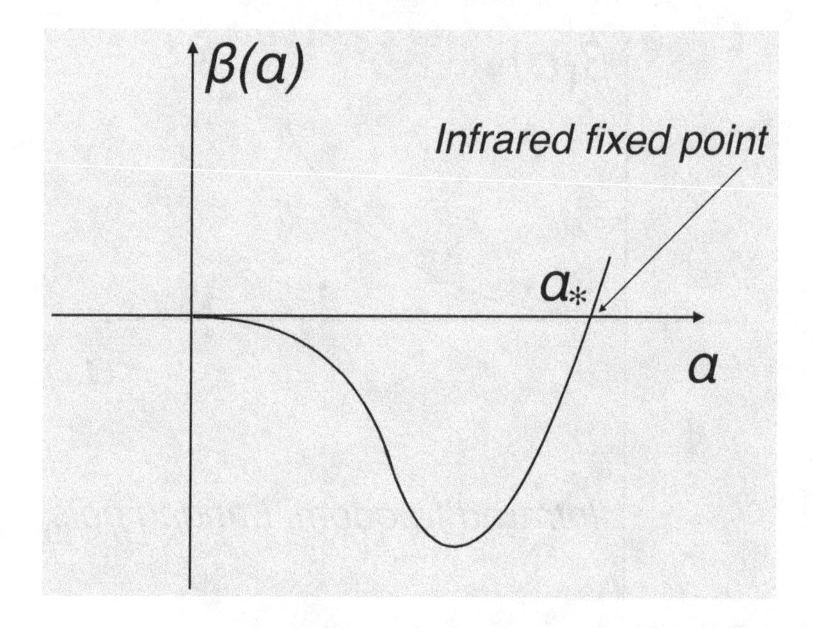

Fig. 14.3 At $\alpha = \alpha_*$ the theory becomes conformal.

Of course, the simplest conformal case is the identically vanishing function $\beta(\alpha)$. This only happens in some (rather rare) supersymmetric field theories.

From pure Yang-Mills theory we pass to theories with matter. Considering N_f massless quarks in the fundamental representation is the first step. Each quark is described by the Dirac spinor, the overall number of the Dirac spinors is N_f. At $N = 3$ and $N_f = 3$ (three light quarks) we obtain quantum chromodynamics (QCD), *the* theory of strong interactions in nature.

The most obvious impact of massless quarks is the change of b, the first coefficient in the Gell-Mann–Low function. We now have

$$b = \frac{11}{3} N - \frac{2}{3} N_f \,. \tag{14.24}$$

If $N_f > \frac{11}{2} N$, the coefficient changes sign, we lose asymptotic freedom, and the Landau regime sets in. The theory becomes infrared free, much like QED with massless electrons. From the point of view of dynamics, this is a rather uninteresting regime.

Let us assume that $N_f \le \frac{11}{2} N$. Now we ask ourselves the question what happens if N_f is only slightly less than the critical value $\frac{11}{2} N$. To answer this question we will need to know the two-loop coefficient in the β function. This will be considered in Sec. 15.3.

Appendix 14.1: Gauge coupling renormalization in gauge theories. Screening versus antiscreening *

It is instructive to consider the gauge coupling (charge) renormalization in the ghost-free Coulomb gauge. Our task is to compare Abelian vs. non-Abelian gauge theories.[3]

Let us start from quantum electrodynamics (QED). Assume we have two heavy (probe) charged bodies, with the charges $\pm e_0$ where $\pm e_0$ is a bare electric charge appearing in the Lagrangian. One can measure the charge through the Coulomb interaction of the probe bodies. The corresponding Feynman diagram is shown in Fig. 14.4 where the wavy line depicts the photon exchange. The heavy probe bodies are (almost) at rest; the photon four-momentum q is assumed to tend to zero. We will choose the reference frame in which $q^\mu = \{q_0,\, 0,\, 0,\, q^3\}$. If $\Gamma_\mu^{(1)}$ and $\Gamma_\mu^{(2)}$ are the vertices for the first and second probe bodies, the amplitude $\mathcal{A}$ corresponding to Fig. 14.4 can be written as

$$\mathcal{A}_0 = \frac{e_0^2}{q^2}\,\Gamma_\mu^{(1)}\,\Gamma_\nu^{(2)}\,g^{\mu\nu} \tag{14.25}$$

where I took into account transversality of the vertices,

$$q^\mu\,\Gamma_\mu^{(1)} = q^\mu\,\Gamma_\mu^{(2)} = 0\,, \tag{14.26}$$

and the subscript 0 (in $\mathcal{A}_0$) means that we deal with the tree diagram (no loops).

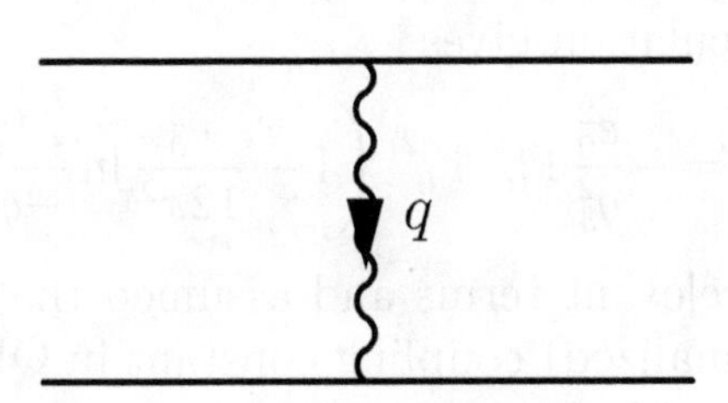

Fig. 14.4 Scattering of two heavy probe charges (denoted by thick lines) in QED in the tree approximation. The photon exchange is denoted by a wavy line. The momentum transfer is q.

The very same transversality implies that $q_0\,\Gamma_0^{(1,2)} = q_3\,\Gamma_3^{(1,2)}$. Using

[3] In this Appendix, I follow Ref. [4]. It is complementary and can be omitted in the first reading.

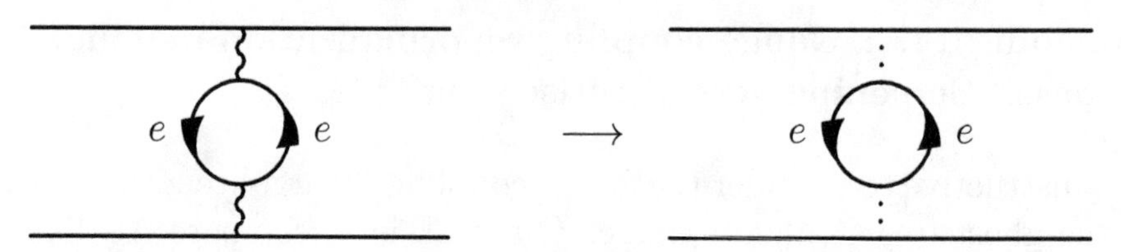

Fig. 14.5 One-loop correction to the Coulomb interaction in QED. The Coulomb part of the photon propagator D_{00} is denoted by the dotted line.

these conditions in Eq. (14.25) we arrive at

$$\mathcal{A}_0 = \frac{e_0^2}{q^2}\left\{\Gamma_0^{(1)}\Gamma_0^{(2)}\left(1-\frac{q_0^2}{q_3^2}\right) - \sum_{\ell=1,2}\Gamma_\ell^{(1)}\Gamma_\ell^{(2)}\right\}$$

$$= -e_0^2\left\{\frac{1}{q_3^2}\Gamma_0^{(1)}\Gamma_0^{(2)} + \frac{1}{q^2}\sum_{\ell=1,2}\Gamma_\ell^{(1)}\Gamma_\ell^{(2)}\right\}. \qquad (14.27)$$

The first term in the braces describes instantaneous Coulomb interaction (this is obvious upon performing the Fourier transformation and passing to the coordinate space). The second term has a pole at $q^2 = 0$. It describes a (retarded) propagation of the electromagnetic wave with two transverse polarizations. We determine the charge through the measurement of the Coulomb interaction. Thus, for our purposes the second term in the braces can be omitted.

The one-loop correction to the Coulomb interaction (14.27) in QED is given by the diagram in Fig. 14.5, with the electron in the loop. A straightforward calculation gives

$$(\mathcal{A}_0 + \mathcal{A}_1)_{\rm QED} = -\frac{e_0^2}{q_3^2}\Gamma_0^{(1)}\Gamma_0^{(2)}\left(1 - \frac{e_0^2}{12\pi^2}\ln\frac{M_{\rm uv}^2}{-q^2}\right) + \ldots \qquad (14.28)$$

where I omitted irrelevant terms and assumed that $|q^2| \gg m_e^2$. Thus, the effective (renormalized) coupling constant in QED which measures the strength of interaction at the scale q^2 (note, that in process at hand $-q^2 < 0$) is

$$e^2(q^2) = e_0^2\left(1 - \frac{e_0^2}{12\pi^2}\ln\frac{M_{\rm uv}^2}{-q^2}\right)$$

$$\to \frac{e_0^2}{1 + \frac{e_0^2}{12\pi^2}\ln\frac{M_{\rm uv}^2}{-q^2}} \qquad (14.29)$$

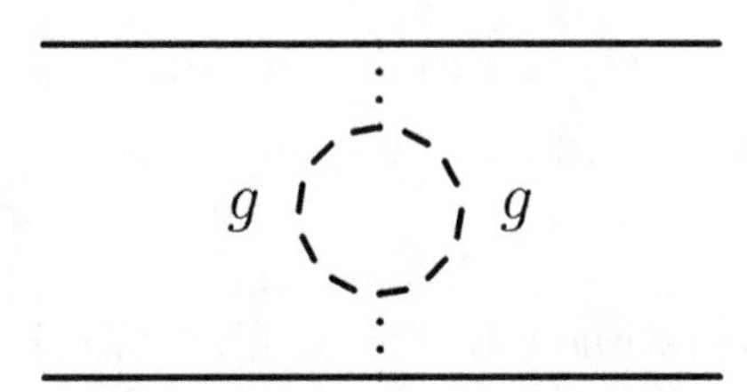

Fig. 14.6 One-loop correction to the Coulomb interaction in Yang-Mills theory. The transverse (physical) gluons are denoted by dashed lines. This diagram is similar to that in Fig. 14.5.

where the first line presents the one-loop expression while the second line is the result of summing up all leading logarithms (which can be performed, e.g. via the renormalization group). At the scale q^2, the effective charge is smaller than the bare one. This is natural; the reason is obvious: the bare charge screening. Indeed, the bare charge is defined at the shortest distances $\sim M_{\rm uv}^{-1}$. Assume for definiteness, the probe charge is positive. Electron-positron pairs created in the vacuum as a result of field-theoretic fluctuations polarize the vacuum. The probe charge attracts negatively-charged electrons while positively-charged positrons are repelled. Then a cloud of virtual electrons screens the original positive probe charge. An effective charge seen at some distance from the probe is smaller than the bare charge, and the further we go, the smaller the screened charge is.

The bare charge screening is a rather general phenomenon. It takes place in all four-dimensional renormalizable field theories except non-Abelian Yang-Mills theories. Formally, the bare charge screening is in one-to-one correspondence with the fact that the imaginary part of the diagram in Fig. 14.5 (the discontinuity in q^2 at positive q^2) is always positive, due to unitarity.

One can ask then: what miracle happens in passing from QED to non-Abelian Yang-Mills theories? In QED, the photons are coupled to the electrons and do not interact with each other directly. In non-Abelian Yang-Mills theories gluons themselves are the sources for gluons (gauge bosons). There are free and four-gluon vertices. Moreover, as we know from Eq. (14.27), the gluon "quanta" can be Coulomb (its propagation is described by the component D_{00} of the Green function) and physical transversal (their propagation is described by the component $D_{\ell\ell'}$ of the Green function, $\ell,\ \ell' = 1, 2$). The diagram depicted in Fig. 14.6 is qualitatively similar to the one-loop correction in QED in Fig. 14.5(b). It produces screening of the bare charge. The only difference with Eq. (14.27) is an insignificant numerical difference: the coefficient $-\frac{e_0^2}{12\pi^2}$ is replaced by $-\frac{g_0^2}{24\pi^2}$ in the SU(2) Yang-Mills theory.

A qualitative difference arises due to the diagram in Fig. 14.7. This graph depicts the transition of the Coulomb quantum (described by D_{00}) into a pair "transverse plus Coulomb." We should remember that the Coulomb interaction is instantaneous: D_{00} depends on q_3^2 rather than on q^2, see (14.27). This means that the contribution of this diagram does not have imaginary part (no discontinuity in q^2 at positive q^2). Unitarity no longer determines the sign of this correction. An explicit calculation shows that it has the opposite sign; the graph in Fig. 14.7 produces antiscreening,

$$(\mathcal{A}_0 + \mathcal{A}_1)_{\text{SU(2) YM}} = \mathcal{A}_0 \left\{ 1 - \frac{g_0^2}{16\pi^2} \frac{2}{3} \ln \frac{M_{\text{uv}}^2}{-q^2} + \frac{g_0^2}{16\pi^2} 8 \ln \frac{M_{\text{uv}}^2}{q_3^2} \right\},$$

$$\to \frac{\mathcal{A}_0}{1 - \frac{g_0^2}{16\pi^2} \ln \frac{M_{\text{uv}}^2}{-q^2} \left(8 - \frac{2}{3}\right) \ln \frac{M_{\text{uv}}^2}{q_3^2}}, \tag{14.30}$$

where the first and the second corrections in the braces are due to Figs. 14.6 and 14.7, respectively.[4] Now the bare charge is smaller than that seen at a distance!

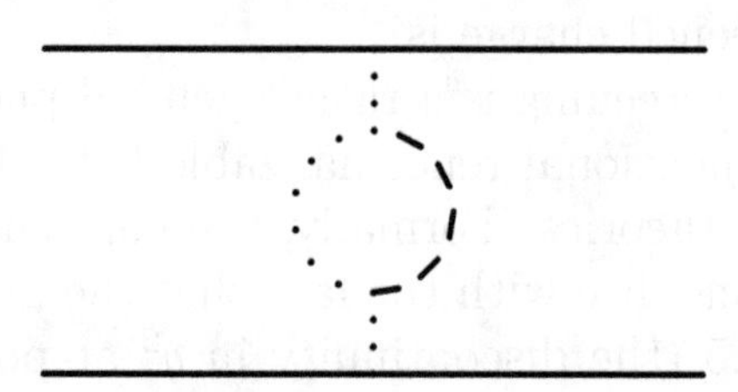

Fig. 14.7 One-loop correction to the Coulomb interaction specific to non-Abelian Yang-Mills theories.

One can give an heuristic argument why these two diagrams produce effects of opposite sign. To this end let us compare the loops in these graphs, as in Fig. 14.8, where I cut one transverse gluon line in order to further clarify an analogy with QED to be presented momentarily. Figure 14.8(a) contains an exchange of the Coulomb quantum,

[4]The result presented in the first line, precisely in this form, was obtained by I. Khriplovich [5] before the discovery of asymptotic freedom and the advent of QCD. A curious story of "pre-observation" of asymptotic freedom is recounted in [6].

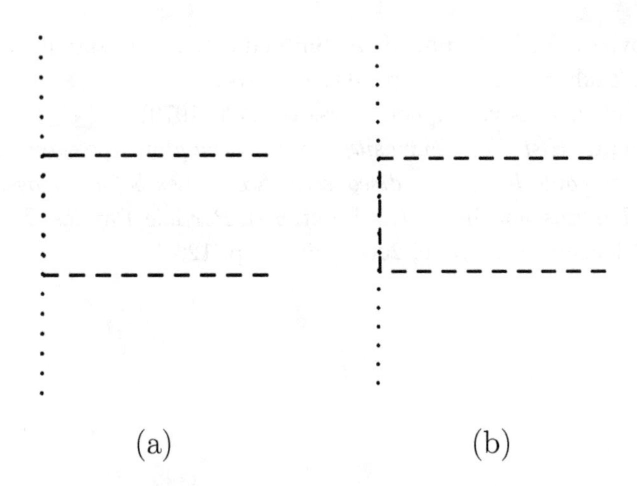

Fig. 14.8 Comparison of loops in Figs. 14.6 and 14.7. Interaction via exchange of the (a) Coulomb and (b) transverse quanta.

while Fig. 14.8(b) transverse gluon quantum. The effect of the Coulomb quanta results in repulsion of charges of the same sign, while the exchange of transverse quanta leads to an attraction of parallel currents (the Biot–Savart law).

The only circumstance which remains unexplained by the above arguments is why the antiscreening represented by the coefficient 8 in Eq. (14.30) is numerically so much stronger in Yang-Mills than the screening effect represented by $2/3$. This is a lucky circumstance since the numerical dominance of antiscreening over screening makes non-Abelian Yang-Mills theories asymptotically free.

It is remarkable that the same binary fission of the one-loop quantum correction, eight vs. $-2/3$, is clearly seen in the background field calculation, see Chapter 11, Eq. (11.27).

References

[1] K. G. Wilson, *Nonlagrangian Models of Current Algebra*, Phys. Rev. **179**, 1499 (1969); for an adaptation to QCD see V. A. Novikov, M. A. Shifman, A. I. Vainshtein and V. I. Zakharov, *Wilson's Operator Expansion: Can It Fail?*, Nucl. Phys. B **249**, 445 (1985).

[2] D. J. Gross and F. Wilczek, Phys. Rev. Lett. **30**, 1343 (1973); H. D. Politzer, Phys. Rev. Lett. **30**, 1346 (1973).

[3] N. Seiberg and E. Witten, Nucl. Phys. B **426**, 19 (1994) [Erratum-ibid. B **430**, 485 (1994)] [arXiv:hep-th/9407087].

[4] V. A. Novikov, L. B. Okun, M. A. Shifman, A. I. Vainshtein, M. B. Voloshin and V. I. Zakharov, Phys. Rept. **41**, 1 (1978).

[5] I. B. Khriplovich, Sov. J. Nucl. Phys. **10**, 235 (1970).

[6] M. Shifman, *Historical curiosity: How asymptotic freedom of the Yang-Mills theory could have been discovered three times before Gross, Wilczek and Politzer, but was not*, in *At the Frontier of Particle Physics*, Ed. M. Shifman (World Scientific, Singapore, 2000), Vol. 1, p. 126.

Chapter 15

Lecture 15. Phases of Yang-Mills Theories

The phase structure of non-Abelian gauge theories is richer than that of QED. In addition to three regimes we had already discussed (Coulomb, Higgs, asymptotically free fixed point), which were known already in the 1970s, Yang-Mills and other theories can exhibit confining and conformal phases, phases with or without chiral symmetry breaking, and so on.

One should not think that the Coulomb phase is a feature of only QED. In Problem 4 (pages 250 and 251) we explored the Georgi-Glashow model, SU(2) Yang-Mills theory with the adjoint Higgs field. At distances $< (gv)^{-1}$ it exhibits asymptotic freedom, while at large distances, $L \gg (gv)^{-1}$, it is realized in the Coulomb phase. This is an interesting example since both the Higgs and Coulomb behaviors are envolved.

The zeroes of the β functions, such as in Fig. 14.3, appear in many theories used in condensed matter to describe phase transition. At $\mu \to 0$ the running of the coupling freezes at $\alpha = \alpha_*$ and all scales $\mu \ll \mu_0$ become equally important (see Eq. (14.23)). Physics no longer depends on the scale. At $\alpha = \alpha_*$, the theory becomes scale invariant, which almost always implies the full conformal invariance. The most well-known example of such a phase transition is provided by the Ising model (see Sec. 13.4). At a certain temperature $T = T_c$ this model becomes conformal and reduces to a free Majorana fermion.

After this brief introduction, I will focus on the phases of Yang-Mills theories such as confining, chiral symmetry breaking and conformal. At the very end, I will add an explanatory remark on the Higgs-confinement phase characterized by two regimes connected by a crossover.

15.1 Confinement

We are used to interactions which are characterized by potentials falling off at large distances. This allows us to isolate an object from another object, say, I can take this marker, put it in my pocket and carry it as far away from this whiteboard as I wish. This is not the case in confining theories, in which the interaction energy grows linearly with distance up to indefinitely large distances; the linear growth *never stops.*

Before the discovery of QCD confinement in the 1970s, the only known natural phenomenon which resembles it was the formation of the magnetic flux Abrikosov vortices in superconductors. In fact, these vortices would confine magnetic monopoles if they existed. Alas... they are not found yet.

Now, consider pure Yang-Mills theory (Chapter 11), where the gauge group is assumed to be SU(N) with the arbitrary value of N. At short distances, the running coupling constant falls off logarithmically [1],

$$\frac{\alpha(p)}{2\pi}=\frac{1}{b\ln\left(p/\Lambda\right)},\qquad b=\frac{11\,N}{3}\,. \tag{15.1}$$

The above formula was discussed in Chapter 16, Eq. (16.25). Interaction switches off at large p, and one can "detect" – albeit indirectly – the gluon degrees of freedom as they are described by the Lagrangian (1.33).

At large distances (i.e. when p approaches Λ from above) we enter a strong coupling regime. The physically observed spectrum is drastically different from what we see in the Lagrangian (1.33). In the case at hand, an experimentalist, if he or she could exist in the world of pure Yang-Mills, would observe a spectrum of color-singlet glueballs, generally speaking, nondegenerate in masses. One can visualize them as a closed string (or, better to say, a tube), in a highly quantum state, i.e. a string-like field configuration which wildly oscillates, pulsates and vibrates, Fig. 15.1. If we add non-dynamical (very heavy) quarks in the theory, and put the quark and antiquark at a large distance from each other, such a string will stretch between them [1] in an inseparable configuration. What is depicted in this figure is a highly quantum (presumably, nonperturbative) open string configuration with quarks attached at the ends. Trying to push the quarks apart, we just make

[1]The probe quarks Q are those for which pair production in the vacuum can be ignored. This can be achieved by endowing them with a mass $m_Q\to\infty$. As opposed to probe quarks, dynamical quarks q are either massless or light, $m_q\ll\Lambda$.

Fig. 15.1 A quantum closed string as a glueball.

the string longer, while the energy of the configuration grows linearly with separation.

This phase of the theory whose existence was conjectured in 1973 [1] is referred to as color confinement. Although there is no analytic proof of color confinement which could be considered exhaustive, there is ample evidence that this regime does, indeed, take place. First, a version of color confinement was observed in certain supersymmetric Yang-Mills theories [2]. Second, the formation of tube-like configurations connecting heavy probe quarks was demonstrated numerically, in lattice simulations. Here, I will not dwell on the dynamics leading to color confinement.[2] It is worth noting, however, that there are distinct versions of confinement regimes, such as oblique confinement [5], Abelian versus non-Abelian confinement (both in Yang-Mills theories), etc.

Kenneth Wilson was the first to suggest [6] a very convenient criterion indicating whether or not a given gauge theory is in the confinement phase. Consider a gauge theory in the Euclidean space-time. Introduce a close contour as shown in Fig. 15.2. Assume that $T \gg L \gg \Lambda^{-1}$, i.e. the contour is large.[3] Consider the Wilson operator

$$W(C) = \frac{1}{\dim_R} \operatorname{Tr} P\exp\left[i \oint_C A_\mu^a(x)\, T_R^a\, dx\right], \tag{15.2}$$

where the subscript R marks the representation of the gauge group to which the probe quark belongs (usually, it is the fundamental representation). P in front of the integral means path ordering. Quite often, this symbol is omitted in the literature.

The asymptotic form of the vacuum expectation value of $W(C)$ is

$$\langle W(C)\rangle_{\rm vac} \propto \exp\left[-\left(\mu P + \sigma A\right)\right], \tag{15.3}$$

[2] For a discussion of this topic see e.g. my textbook [3] or the review [4].

[3] Generally speaking, the contour does not have to be rectangular, but for the rectangular contour the result is simpler to interpret.

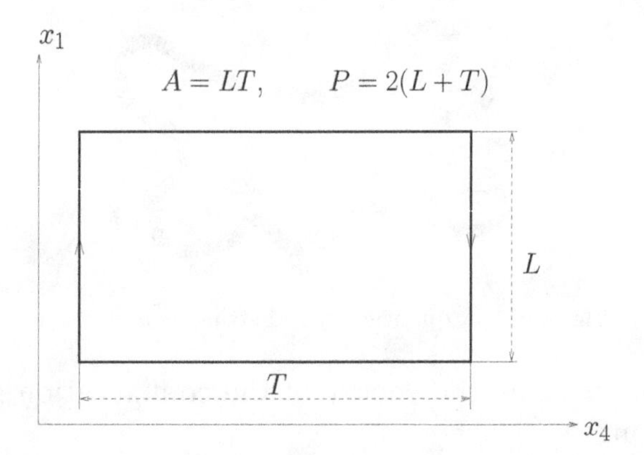

Fig. 15.2 The Wilson contour C, with area A and perimeter P. The probe quark is dragged along this contour.

where $A = LT$ is the area of the contour, $P = 2(L+T)$ – perimeter; μ and σ are numerical coefficients of dimension mass and mass squared, respectively. If

$$\sigma \neq 0, \tag{15.4}$$

the theory is in the confinement phase, while at $\sigma = 0$ it does not confine.[4] We refer to these cases as the area law and perimeter law, respectively.

Why does the area law imply confinement? This is because, on general grounds,

$$\langle W(C)\rangle_{\text{vac}} \propto \exp\left(-V(L)T\right) \tag{15.5}$$

if the contour is chosen as shown above. Hence, the area law means that the potential $V(L)$ between distant probe quarks Q and $\bar{Q}$ is $V(L) = \sigma L$ at $L \gg \Lambda^{-1}$. The coefficient σ is the string tension. (By the way, in many publications it is denoted by T rather than σ.)

15.2 Adding massless quarks

From pure Yang-Mills theory we pass to theories with matter. Considering N_f massless quarks in the fundamental representation is the first step. Each quark is described by the Dirac spinor, the overall number of the Dirac spinors is N_f. At $N = 3$ and $N_f = 3$ we obtain quantum chromodynamics (QCD), *the* theory of strong interactions in nature.

[4]If $\sigma \neq 0$, the perimeter term is subleading. The parameter μ renormalizes the probe quark mass.

The most obvious impact of massless quarks is the change of β_0, the first coefficient in the Gell-Mann–Low function. Instead of (15.1) we now have

$$\beta_0 = \frac{11}{3}\,N - \frac{2}{3}\,N_f\,. \tag{15.6}$$

If $N_f > \frac{11}{2}N$, the coefficient changes sign, we lose asymptotic freedom, and the Landau regime sets in. The theory becomes infrared free, much like QED with massless electrons. From the dynamics standpoint, this is a rather uninteresting regime.

Let us assume that $N_f \leq \frac{11}{2}N$. Now we will address the question on what happens if N_f is only slightly less than the critical value $\frac{11}{2}N$. To answer this question we will need to know the two-loop coefficient in the β function.

15.3 Conformal phase

The response of Yang-Mills theories under the scale and conformal transformations is determined by the trace of the energy-momentum tensor,[5]

$$T^\mu_\mu \propto \beta(\alpha)\, G^a_{\mu\nu}\, G^{\mu\nu,\,a} \tag{15.7}$$

where $\beta(\alpha)$ is the Gell-Mann–Low function (a.k.a the β function). In SU(N) Yang-Mills theory with N_f quarks it has the form

$$\beta(\alpha) = \frac{\partial\alpha(\mu)}{\partial\ln\mu} = -\beta_0\frac{\alpha^2}{2\pi} - \beta_1\frac{\alpha^3}{4\pi^2} - \dots\,, \qquad \alpha = \frac{g^2}{4\pi}\,, \tag{15.8}$$

where β_0 is given in (15.6) while

$$\beta_1 = \frac{17}{3}\,N^2 - \frac{N_f}{6N}\left(13\,N^2 - 3\right)\,. \tag{15.9}$$

At small α the first coefficient in (15.8) dominates, the β function is negative, implying asymptotic freedom at short distances. What is the large-distance behavior of the running coupling constant?

Assume that

$$N_f = \frac{11}{2}\,N - \nu\,, \qquad 0 < \nu \ll \frac{11}{2}\,N\,. \tag{15.10}$$

Then the first coefficient β_0 is anomalously small,

$$\beta_0 = \frac{2}{3}\,\nu\,. \tag{15.11}$$

[5]Equation (15.7) presents the so-called scale anomaly, see Sec. 19.5.

At the same time, the second coefficient is not suppressed, it is of a normal order of magnitude,

$$\beta_1 = -\frac{25}{4}N^2 + \frac{11}{4} + \frac{\nu}{6N}\left(13\,N^2 - 3\right), \tag{15.12}$$

and *negative*!

With the scale μ decreasing (at larger distances) the running gauge coupling constant grows, and the second term eventually becomes important. Generally speaking, the second term takes over the first one at $N\alpha/\pi \sim 1$ (in the strong coupling regime), when *all* terms in the α expansion of the β function are equally important, and by no means one can limit oneself to the first two terms. However, if N_f is only slightly less than $\frac{11}{2}N$ the β function develops a zero at a value of α which is parametrically[6] small, namely,

$$\frac{N\alpha_*}{2\pi} = \frac{N\beta_0}{-\beta_1} = \frac{8}{75}\,\frac{\nu}{Nf(N,\,\nu)} \tag{15.13}$$

where

$$f(N,\,\nu) = 1 - \frac{11}{25\,N^2} - \frac{2\nu}{75\,N^3}\left(13\,N^2 - 3\right) \sim 1\,. \tag{15.14}$$

In other words, the second term catches up with the first one prematurely, when $N\alpha/\pi \ll 1$. Hence, we are at weak coupling, and higher order terms are inessential. The fact of the existence of this zero, as well as its position, are reliably established.

As an example, let me indicate that if $N = 3$ and $N_f = 15$

$$\frac{\alpha_*}{2\pi} = \frac{1}{44}\,. \tag{15.15}$$

A sketch of the β function is given in Fig. 15.3.

The zero of the β function depicted in Fig. 15.3 is nothing but the infrared fixed point of the theory. If we start from the value of α lying between 0 and α_*, and let it run, then the running α will hit α_* in the infrared (remember, in the ultraviolet $\alpha(\mu)$ tends to 0).

Hence, at large distances $\beta(\alpha) = \beta(\alpha^*) = 0$, implying that the trace of the energy-momentum tensor of the theory vanishes, and the theory is in the *conformal* phase, there are no localized particle-like states in the spectrum; rather, we deal with massless unconfined *interacting* quarks and gluons. All correlation functions at large distances exhibit a power-like behavior. As long as α^* is small, the interaction of the

[6] By saying parametrically, I mean, that if, for instance, N is large while ν does not scale with N, then $f(N,\,\nu) \to 1$, and $\frac{N\alpha_*}{2\pi} \to (8/75)(\nu/N)$.

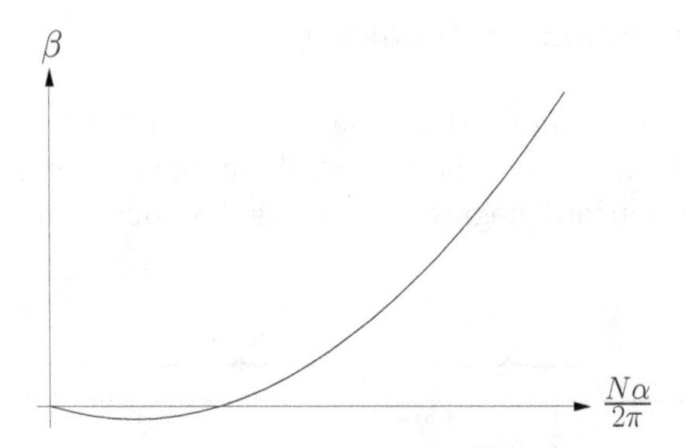

Fig. 15.3 A sketch of the β function at N_f slightly less than $\frac{11}{2}N$. The horizontal axis presents $N\alpha/2\pi$. The zero of the beta function is at $(8/75)(\nu/N) \ll 1$.

massless quarks and gluons in the theory is weak at all distances, short and large, and is amenable to the standard perturbative treatment. In particular, the potential between two probe static quarks at large distance R will approximately behave as α^*/R reminding conventional QED with massive electrons.

Since we are absolutely certain that slightly below $N_f = \frac{11}{2}N$ we are in the conformal phase, increasing ν (decreasing N_f), we cannot leave this phase right away. There should exist a critical value N_f^* of the number of flavors above which the theory is conformal in the infrared. The interval

$$N_f^* \leq N_f \leq \frac{11}{2}N \tag{15.16}$$

is referred to as a *conformal window*.[7] The exact value of N_f^* is unknown. From experiment we know that $N_f^* > 3$ at $N = 3$. On general grounds one can argue that $N_f^* \sim cN$ where c is a numerical constant of the order of unity. Of course, near the left (lower) edge of the conformal window one should expect $\frac{N\alpha_*}{2\pi} \sim 1$, so that the theory, albeit conformal in the infrared, is strongly coupled. In particular, in this case there is no reason for anomalous dimensions to be small.

Summarizing, if N_f lies in the interval (15.16), the theory is in the conformal phase. For N_f close to the right (upper) edge of the conformal window the theory is weakly coupled and all anomalous dimensions are calculable. Belavin and Migdal played with this model in the early 1970s [8]. Somewhat later, this idea was thoroughly studied by Banks and Zaks [9].

[7]This terminology was suggested in [4], and it took root.

15.4 Chiral symmetry breaking

Next, in our journey along the N_f-axis (Fig. 15.4) let us descend down to $N_f = 1, 2, 3, ...$ Strictly speaking, dynamical quarks (in the fundamental representation) negate confinement understood in the sense of

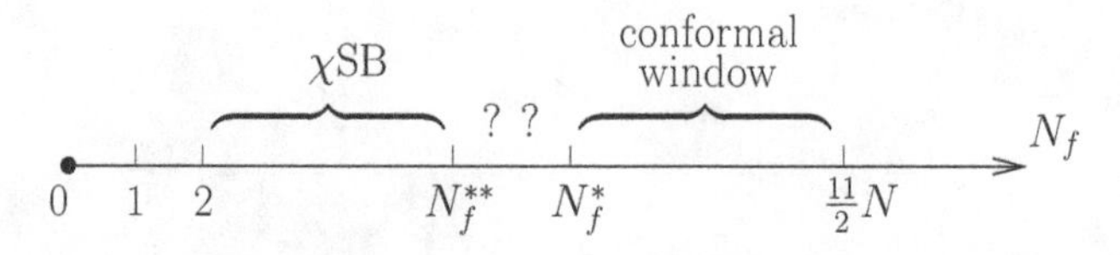

Fig. 15.4 Dynamical regimes change as we change the number of the massless quarks N_f.

Wilson's criterion – the area law for the Wilson loop disappears. Indeed, the string forming between the probe quarks can break through the $\bar{q}q$ pair creation, as long as the energy stored in the string becomes sufficient to produce such a pair (Fig. 15.5). As a result, sufficiently large

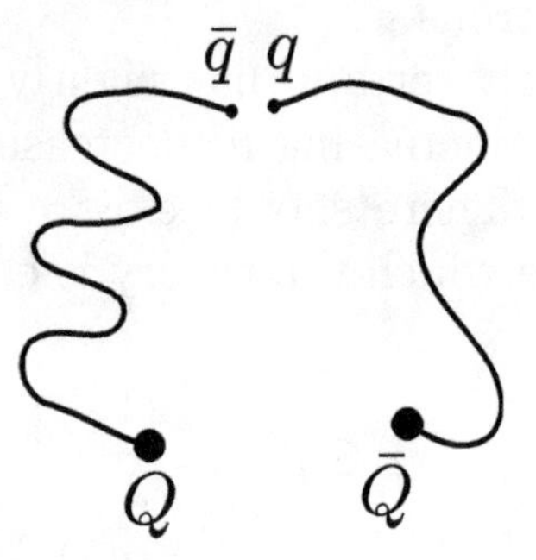

Fig. 15.5 The string between two probe quarks can break through $\bar{q}q$ pair creation in Yang-Mills with dynamical quarks.

Wilson loops obey the perimeter law rather than the area law. However, intuitively it is clear that, in essence, this is the same confinement, although in the case at hand it is natural to call it *quark confinement.* The dynamical quarks are identifiable at short distances in a clear-cut manner; and yet, they never appear as asymptotic states. Experimentalists detect only color-singlet mesons of the type $\bar{q}q$ or baryons of the type qqq.

Theoretically, if necessary, one can suppress the $\bar{q}q$ pair creation by sending $N \to \infty$.

At $N_f \geq 2$, a new interesting phenomenon shows up. The global symmetry of Yang-Mills theory with more than one massless quark flavor is

$$\mathrm{SU}(N_f)_\mathrm{L} \times \mathrm{SU}(N_f)_\mathrm{R} \times \mathrm{U}(1)_V\,. \tag{15.17}$$

The vectorial U(1) symmetry is nothing but the baryon number, while the axial U(1) is anomalous, as we will see later, and, hence, is not shown in (15.17). The origin of the chiral $\mathrm{SU}(N_f)_\mathrm{L} \times \mathrm{SU}(N_f)_\mathrm{R}$ symmetry is as follows. The quark part of the Lagrangian has the form

$$\mathcal{L}_\mathrm{quark} = \sum_f \bar{\Psi}_f\, i \not{D}\, \Psi^f\,, \tag{15.18}$$

where Ψ^f is the Dirac spinor of a given flavor f (the quark mass is neglected). Each Dirac spinor is built of one left- and one right-handed Weyl spinor,

$$\Psi_i^f = \begin{pmatrix} \xi^f_{\alpha,\,i} \\ \bar{\eta}_i^{\dot\alpha,\,f} \end{pmatrix}, \tag{15.19}$$

where i is the color index (i.e. the index of the fundamental representation of $\mathrm{SU}(N)_\mathrm{color}$) while f is the flavor index, $f = 1, 2, 3, \ldots, N_f$. The left- and right-handed Weyl spinors in the kinetic term above totally decouple from each other. Hence, $\mathcal{L}_\mathrm{quark}$ is invariant under independent global rotations

$$\xi \to U\xi\,, \quad \bar\eta \to U'\bar\eta\,, \qquad U \in \mathrm{SU}(N_f)_\mathrm{L}\,, \quad U' \in \mathrm{SU}(N_f)_\mathrm{R}\,. \tag{15.20}$$

Experimentally it is known that the chiral $\mathrm{SU}(N_f)_\mathrm{L} \times \mathrm{SU}(N_f)_\mathrm{R}$ symmetry is spontaneously broken at $N = 3$ and $N_f = 2, 3$. $N_f^2 - 1$ massless Goldstone bosons – the pions [8] – emerge as a result of this spontaneous breaking. This phenomenon bears the name of the *chiral symmetry breaking* (χSB).

There are qualitative arguments showing that in four-dimensional Yang-Mills theory χSB may be a consequence of quark confinement plus some general features of the quark-gluon interaction. In particular, well-known is Casher's qualitative picture [10] "explaining"[9] why in Yang-Mills theories with massless quarks (no scalar fields!) color confinement entails the Goldstone-mode realization of the global axial

[8] The Goldstone mesons are called pions in the case of two quark flavors. For $N_f = 3$ the Goldstone meson family includes, in addition, K mesons and η. This octet is referred to as the Goldstone octet, or "generalized" pions.

[9] I put explaining in the quotation marks since Casher's consideration is rather nebulous and imprecise.

symmetry of the Lagrangian. A brief outline is as follows. If we deal with massless quarks, the left-handed quarks are decoupled from the right-handed quarks in the QCD Lagrangian. If spontaneous breaking of the chiral symmetry does not take place, this decoupling becomes an exact property of the theory: the quark chirality (helicity) is exactly conserved. Assume that we produce an energetic quark-antiquark pair, say, in e^+e^- annihilation. Let us place the origin at the annihilation point. If quarks' energy is high they can be treated quasiclassically. Let us say that in the given event the quark produced is right-handed and flies in the positive z direction; the antiquark will then fly in the negative z direction. If the quark energy is high ($E \gg \Lambda$ where Λ is the QCD scale parameter) the distance L the quark travels before confining effects become critical is large, $L \sim E/\Lambda^2$. Color confinement means that the quark cannot move indefinitely in the positive z direction; at a certain time $T \sim E/\Lambda^2$, it should turn back and start moving in the negative z direction. Consider this turning point in more detail. Before the turn, the quark's spin projection on the z-axis is $+1/2$. Since, by assumption, quark's helicity is conserved, after the turn, when p_z becomes negative, the quark's spin projection on the z-axis must be $-1/2$ (Fig. 15.6). In other words, $\Delta S_z = -1$. The total angular momentum is conserved. Consequently, $\Delta S_z = -1$ must be compensated. At the time of the turn, the quark is far away from the antiquark; they do not "know" what their respective partners do; conservation of the angular momentum must be achieved locally. The only object that could have been responsible for compensation of quark's ΔS_z is a QCD string that stretches in the z direction between the quark and the antiquark. The QCD string is responsible for color confinement, but it does not have L_z (more exactly, it is *presumed* to have no L_z) and, thus, cannot support conservation of the angular momentum in this picture. Thus, either the quark never turns (no confinement) or, if it does, the chiral symmetry *must* be spontaneously broken.

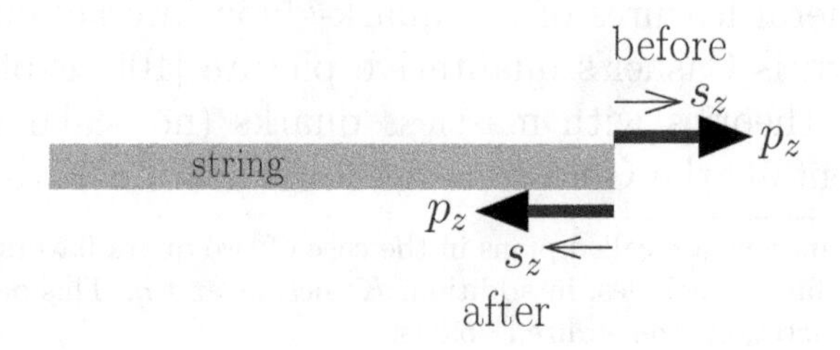

Fig. 15.6 Right-handed quark before and after turning point.

The relation between quark confinement and χSB is a deep and intriguing dynamical question. Since I have nothing to add, let me summarize. There is a phase of QCD in which quark confinement and χSB coexist. On the N_f-axis this phase starts at $N_f = 2$ and extends up to some upper boundary $N_f = N_f^{**}$. We do not know whether or not N_f^{**} coincides with the left edge of the conformal window N_f^*. It may happen that $N_f^{**} < N_f^*$, and the interval $N_f^{**} < N_f < N_f^*$ is populated by some other phase or phases (e.g. confinement without χSB) ...

15.5 A few words on other regimes

Using various ingredients and mixing them in various proportions to construct the matter sector with desired properties, one can reach other phases of Yang-Mills theories. For instance, Higgsing the theory as in Chapter 3 and breaking SU(N) down to U(1)$^{N-1}$ we can implement the Coulomb phase. Let us ask ourselves what happens if this Higgsing is implemented through the scalar fields in the fundamental representation, as in Chapter 3. If $v \gg \Lambda$ the theory is at weak coupling; it resembles the standard model. On the other hand, if $v \ll \Lambda$ the theory is at strong coupling. Our intuition tells us that in this case it should resemble QCD, with a rich spectrum of composite color-singlet mesons, with all possible quantum numbers.

There are convincing arguments [11] that there is no phase transition between these two regimes. Indeed, if the scalar fields are in the fundamental representation, the color-singlet interpolating operators built from these fields, the gluon field strength tensor and covariant derivatives span the space of physical (color-singlet or gauge invariant) states in its *entirety.* All possible quantum numbers are covered. As we change the vacuum expectation value v from small to large, the strong coupling regime smoothly gives way to the weak coupling regime, possibly with a crossover in the middle. Each state existing at strong coupling is mapped onto its counterpartner at weak coupling.

For instance, consider the operator

$$\mathrm{Tr}\left(\bar{X}\, i\overleftrightarrow{\mathcal{D}}_\mu X\, \tau^a\right). \tag{15.21}$$

At $v \ll \Lambda$ this operator produces a "ρ meson" and its excitations. The low-lying excitations could be seen as resonances. As v increases and becomes much larger than Λ, the very same operator obviously reduces to $v^2\, W_\mu +$ small corrections. It produces a W boson from the vacuum.

It produces excitations too, but they are no longer resonances, rather, they are states which contain a number of W bosons and Higgs particles with the overall quantum numbers of a single W boson. Note that the global SU(2) symmetry of the model of Chapter 3 is respected in both regimes. All states come out in complete representations of SU(2), e.g. triplets, octets, and so on.

In the general case, I formulate the following *statement* which, being short of the rigorous mathematical proof, can be accepted by physicists [13] (see also [14]):

If, in addition to gauge fields, a given non-Abelian theory contains a set of Higgs fields in the fundamental representations which, by developing vacuum expectation values (VEVs), can Higgs the gauge group completely, then, decreasing all the above VEVs (proportionally to each other), from large to small values, we do not pass through the Higgs/confinement phase transition. Rather, a crossover from weak to strong coupling takes place.

Contrived matter sectors can lead to more "exotic" phases. I have already mentioned oblique confinement. In supersymmetric Yang-Mills theories with matter in the adjoint representation, a number of unconventional phases was found in [12]. I will not dwell on them as this aspect goes far beyond my course.

References

[1] D. J. Gross and F. Wilczek, Phys. Rev. Lett. **30**, 1343 (1973); H. D. Politzer, Phys. Rev. Lett. **30**, 1346 (1973).

[2] N. Seiberg and E. Witten, Nucl. Phys. B **426**, 19 (1994) [Erratum-ibid. B **430**, 485 (1994)] [arXiv:hep-th/9407087].

[3] M. Shifman, *Advanced Topics in Quantum Field Theory*, (Cambridge University Press, 2012).

[4] M. A. Shifman, Prog. Part. Nucl. Phys. **39**, 1 (1997) [arXiv:hep-th/9704114].

[5] G. 't Hooft, Nucl. Phys. B **190**, 455 (1981).

[6] K. G. Wilson, Phys. Rev. D **10**, 2445 (1974).

[7] Y. Frishman and J. Sonnenschein, *Non-Perturbatice Field Theory*, (Cambridge University Press, 2010).

[8] A. Belavin and A. Migdal, JETP Lett. **19**, 181 (1974); and *Scale Invariance and Bootstrap in the Non-Abelian Gauge Theories*, Landau Institute Preprint-74-0894, 1974 (unpublished).

[9] T. Banks and Zaks, Nucl. Phys. B **196**, 189 (1982).

[10] A. Casher, Phys. Lett. B **83**, 395 (1979).

[11] K. Osterwalder and E. Seiler, Annals Phys. **110**, 440 (1978); T. Banks and E. Rabinovici, Nucl. Phys. B **160**, 349 (1979); E. H. Fradkin and S. H. Shenker, Phys. Rev. D **19**, 3682 (1979).

[12] F. Cachazo, N. Seiberg and E. Witten, JHEP **0302**, 042 (2003) [arXiv:hep-th/0301006].

[13] E. H. Fradkin and S. H. Shenker, Phys. Rev. D **19**, 3682 (1979).

[14] M. Shifman and A. Yung, *Fradkin-Shenker Continuity and "Instead-of-Confinement" Phase*, Mod. Phys. Lett. A **32**, 1750159 (2017).

Chapter 16

Lecture 16. Anomalous Dimensions. Quark Mass in QCD

Anomalous scaling dimensions of local operators in asymptotically free field theories. Briefly on a general case of the UV and IR scaling behavior. A sample calculation.

16.1 Anomalous dimension

Now, when we are acquainted with the notion of the renormalization group flow and β functions we will take a new step and will introduce a related concept of *anomalous dimensions.* Anomalous dimensions describe deviation of the scaling laws (from the canonic laws) of various operators considered as a function of the sliding scale μ in the effective Lagrangian.

In free field theories (i.e. in the limit of vanishing coupling constants) all operators have normal dimension, i.e. they scale in accordance with their canonic mass dimensions, and so do various parameters. For instance, if we scale

$$x \to \lambda^{-1} x, \quad m \to \lambda m \tag{16.1}$$

then the gluon field strength tensor will scale as $G_{\mu\nu} \to \lambda^2 G_{\mu\nu}$, the vector current $(\bar{\psi}\gamma_\mu\psi) \to \lambda^3(\bar{\psi}\gamma_\mu\psi)$, the scalar density $(\bar{\psi}\psi) \to \lambda^3(\bar{\psi}\psi)$, and so on. The powers of λ give the so-called *normal (or canonic) scaling dimensions.*

However, in interacting field theories the actual scaling dimensions can deviate from the normal dimensions. The deviation is referred to as the anomalous scaling dimensions. One should distinguish the scaling dimensions in UV from those in IR. For instance, the famous scaling dimensions in the Ising model (Chapter 13) refer to temperatures near the phase transition and correspond to the scaling behavior in the conformal (infrared) limit of this theory. They cannot be calculated perturbatively. This is a rather typical situation in condensed matter physics.

In this lecture, we will consider the scaling dimensions of asymptotically free field theories in the limit $\mu \gg \Lambda$ and $\alpha \ll 1$. This is obviously the UV limit. As we already know, the point $\alpha = 0$ is a zero of the β function. This is a special zero, however. At this point, the β function touches the zero axis rather than crosses it. Its derivative $\partial\beta/\partial\alpha$ vanishes at $\alpha = 0$. Because of this, the approach to the limit $\alpha = 0$ is very slow, logarithmic. The same will be valid for the UV anomalous dimensions – they will depend on the scale μ (and, hence, the parameter λ) *logarithmically.* They can be calculated in perturbation theory. Below, I will consider one of the most important examples, QCD. Moreover, I will focus just on a single operator, the scalar density, and the mass parameter which comes with it.

In this chapter the gauge coupling constant will be denoted as α_S, with the subscript S appearing in QCD for historical reasons.

We pass from the general formalism (Eqs. (14.7) and (14.8) in Chapter 14) to this particular example,

$$\mathcal{L} = -\frac{1}{4g^2} G^a_{\mu\nu}\, G^{\mu\nu\, a} + \bar\Psi\,(iD_\mu\gamma^\mu\Psi) - m\bar\Psi\Psi\,. \tag{16.2}$$

The last term will be considered as “small”, i.e. we will keep it only in the $O(m)$ approximation and will not iterate in m. This is the case in which the notion of the anomalous dimension is applicable. We will limit ourselves to one loop (more exactly, to the leading logarithmic approximation, LLA, which is fully determined by one loop).

Renormalization of $G^a_{\mu\nu}\, G^{\mu\nu\, a}$ and the running of the gauge coupling were discussed in Chapter 11. Now we will examine what happens with the last two terms in (16.2) in the process of evolution from UV at M_0 to a lower point μ.[1] It will be assumed that

$$\mu \gg |m|\,. \tag{16.3}$$

The external fermion line carries momentum q, while the gluon momentum in the loop is p, see Fig. 16.1. For simplicity we will choose q as follows:

$$\mu \gg q\,, \quad |q| \gg |m|\,. \tag{16.4}$$

[1]The scalings in μ is equivalent to that in λ, see Sec. 16.2.1 below.

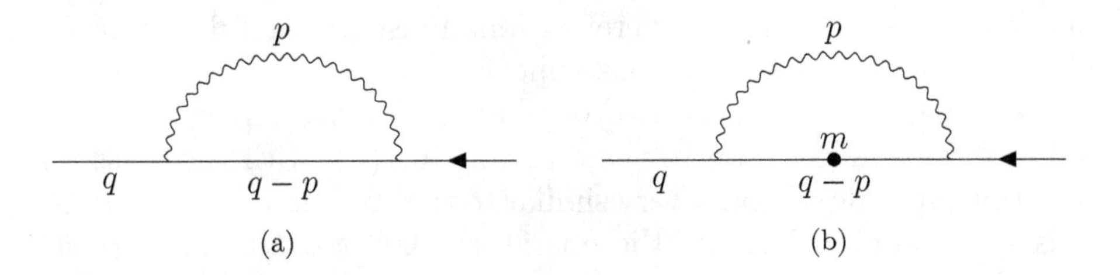

Fig. 16.1 Relevant Feynman graphs: (a) Z factor; (b) mass anomalous dimension.

It is quite clear that the only modification in the last two terms in (16.2) that can happen under evolution is the occurrence of Z factors depending on the gluon coupling g^2 and the ratio M_0^2/μ^2,

$$\Delta\mathcal{L} = Z_\Psi\,\bar\Psi\,(iD_\mu\gamma^\mu\Psi) - Z\,m\bar\Psi\Psi\,. \tag{16.5}$$

For historical reasons the first Z factor, Z_Ψ, is usually referred to as the fermion wave function renormalization. In more modern literature people tend to use the terminology in which Z factors are labeled by subscripts.

Z_Ψ is easy to calculate from the graph of Fig. 16.1(a). Indeed, we start from

$$\mathcal{L}_\Psi = \bar\Psi\,(iD_\mu\gamma^\mu\Psi) \tag{16.6}$$

at M_0 and then integrate out all virtual momenta from M_0 down to μ. The Feynman graph 16.1(a) yields (in the Feynman gauge, important!)

$$\begin{aligned}\Delta\mathcal{L}_\Psi &= ig^2\left(\bar\Psi\gamma_\mu\,T^a\,\frac{-i}{p^2}\,\frac{i(q_\alpha-p_\alpha)\gamma^\alpha}{(q-p)^2}\,T^a\,\gamma^\mu\,\Psi\right)\\ &= \frac{-8ig^2}{3}\,\bar\Psi\left[\frac{q_\alpha\gamma^\alpha}{p^4}-\frac{p_\alpha\gamma^\alpha}{p^4}\left(1+\frac{2pq}{p^2}\right)\right]\Psi\\ &= \frac{4g^2}{3}\left(\int\frac{d^4p}{p^4}\right)_{\rm Eucl.}\bar\Psi q_\alpha\gamma^\alpha\Psi\,.\end{aligned} \tag{16.7}$$

Equation (16.7) implies that the Z_Ψ factor is

$$Z_\Psi = 1+\frac{4}{3}\,\frac{g^2}{16\pi^2}\log\frac{M_0^2}{\mu^2}\,. \tag{16.8}$$

Note that Z_Ψ is gauge dependent if the fermions are off mass shell (and they are). However, the ratio of Z/Z_Ψ is gauge independent since it determines the physically observable effect: the RG evolution of the quark mass (see (16.11)). Therefore, one must be careful and perform all calculations of the anomalous dimension in one and the same gauge. The calculation above is carried out in the *Feynman gauge.*

Now, we will move on to the diagram 16.1(b), with the mass term insertion. Its calculation is very similar to that presented above. In fact, it is even simpler, because the operator $m\bar{\Psi}\Psi$ contains no derivative and, therefore, with the logarithmic accuracy one can put $q = 0$ in the integrand. Then

$$m\bar{\Psi}\Psi\Big|_{M_0} \to \left(1 + \frac{16g^2}{3}\int \frac{d^4p}{p^4}\Big|_{\text{Euclid}}\right)\left(\bar{\Psi}m\Psi\right)_{M_0}$$

$$= \left(1 + \frac{16}{3}\,\frac{g^2}{16\pi^2}\log\frac{M_0^2}{\mu^2}\right)\left(\bar{\Psi}m\Psi\right)_{M_0} \tag{16.9}$$

The expression in the large parentheses on the second line gives us the Z factor.

Next, we return to Eq. (16.5). The original definition of the mass term refers to the Lagrangian with canonically normalized kinetic term. Hence, after the evolution we must redefine the Ψ fields to make the kinetic term canonic. To this end, we introduce new fields $\Psi(\mu)$ such that $\sqrt{Z_\Psi}\Psi_0 = \Psi(\mu)$. Then Eq. (16.5) takes the form

$$\Delta\mathcal{L} = \bar{\Psi}\left(iD_\mu\gamma^\mu\Psi\right)_\mu - \frac{Z}{Z_\Psi}\,m_0\left(\bar{\Psi}\Psi\right)_\mu . \tag{16.10}$$

If we denote

$$\frac{Z}{Z_\Psi}\,m_0 \equiv m(\mu) \tag{16.11}$$

the mass term $\bar{\Psi}m\Psi$ is not renormalized, i.e. its expression in terms of the UV quantities is the same as in terms of the running quantities.

Equation (16.11) implies that

$$m(\mu) = \left(1 + 4\,\frac{g^2}{16\pi^2}\log\frac{M_0^2}{\mu^2}\right)m_0 . \tag{16.12}$$

The anomalous dimension is introduced in parallel with the β functions, namely,[2]

$$\mu\,\frac{d}{d\mu}\log m(\mu) = \gamma_m(\alpha_S(\mu)) . \tag{16.13}$$

[2]Explain why the right-hand side of (16.13) is m-independent. Hint: take into account (16.4). In QCD it is customary to denote the coupling $g^2/4\pi$ by α_S. The subscript S reminds us that QCD is the theory of strong interactions.

Let me mention in passing that Eq. (16.13) can be viewed as a version of the so-called Callan-Symanzik equations which are often used in renormalization group studies. I will not discuss them since I want to avoid excessive technicalities in this course, focussing on basics. The reader who needs further details is referred to Chapter 12 of Peskin and Schroeder.

Combining (16.12) and (16.13) we arrive at

$$\gamma_m = -4\frac{\alpha_S}{2\pi}\,. \tag{16.14}$$

One can integrate Eq. (16.13) and obtain in this way the full expression for $m(\mu)$ in LLA, rather than the one-loop expression. Indeed, from Eq. (16.13), we conclude that

$$\log m(\mu)\Big|_{\mu}^{M_0} = \int_{\mu}^{M_0} \gamma(\alpha_S(\mu))d\log\mu$$

$$= \int \frac{\gamma(\alpha_S)}{\beta(\alpha_s)}\,d\alpha_S\,, \tag{16.15}$$

where I used Eq. (14.11) from Chapter 14. The lower limit of integration in the second line is $\alpha_S(\mu)$. Using Eq. (14.15) we obtain our final expression,

$$\log\frac{m_0}{m(\mu)} = \frac{4}{b}\log\frac{\alpha_{S0}}{\alpha_S(\mu)} \tag{16.16}$$

or

$$m(\mu) = m_0\cdot\left[\frac{\alpha_S(\mu)}{\alpha_{S0}}\right]^{4/b}. \tag{16.17}$$

The smaller the value of μ the larger is the running mass, and *vice versa*. In this sense one can say that the mass term exhibits the asymptotically free behavior. The expression in the square brackets in (16.17) is the anomalous dimension of mass.

16.2 More on dimensional transmutation

In Chapter 14 we introduced the running coupling constant in the Yang-Mills theories and derived a law which governs its running,

$$\mu\frac{d\alpha_S}{d\mu} = \beta(\alpha_S)\,. \tag{16.18}$$

where

$$\alpha_S(\mu) = \frac{g^2(\mu)}{4\pi} \, . \tag{16.19}$$

We calculated the one-loop expression for $\alpha_S(\mu)$,

$$\alpha_S(\mu) = \frac{\alpha_0}{1 - b\,(\alpha_0/4\pi)\log M_0^2/\mu^2} \, , \tag{16.20}$$

where the parameter b is

$$b = \frac{11}{3}N - \frac{2}{3}n_f - \frac{1}{6}n_s \, , \tag{16.21}$$

and then established that in LLA

$$\mu\,\frac{d\alpha_S}{d\mu} = \beta(\alpha_S) = -b\,\frac{\alpha_S^2}{2\pi} \, . \tag{16.22}$$

Let us introduce a parameter Λ (referred to as a dynamical scale) such that formally at $\mu = \Lambda$ the denominator in (16.20) vanishes and α_S explodes,

$$1 = b\frac{\alpha_0}{2\pi}\log\frac{M_0}{\Lambda} \, . \tag{16.23}$$

Equation (16.23) implies that

$$\Lambda = M_0 \exp\left(-\frac{2\pi}{b\alpha_0}\right) \, . \tag{16.24}$$

Thus, we trade the unknown UV data (in a special combination, see (16.26)) in favor of a parameter which sets the scale for all physically measurable phenomena. This is called *dimensional transmutation.* The origin of this term is as follows. In the Lagrangian, we have only a dimensionless parameter α_0. Instead, physics is determined by a dimensionful parameter Λ.

Substituting (16.23) in denominator of (16.20) we arrive at

$$\alpha_S(\mu) = \frac{2\pi}{b\,\log\left(\mu/\Lambda\right)} \, , \tag{16.25}$$

We see that the physical coupling constant $\alpha_S(\mu)$ at a physical scale μ depends only on the dynamical scale Λ. As μ increases, the corresponding coupling constant $\alpha_S(\mu)$ decreases. This is called *asymptotic freedom*, and it occurs only in Yang-Mills theory. In QED and other 4D field theories the law of running is opposite. This is regulated by the sign of the β function: for asymptotic freedom we need it to be negative.

Needless to say, Eq. (16.25) is valid only for $\mu \gg \Lambda$; if $\mu \sim \Lambda$ perturbation theory does not work. Equation (16.25) can be rewritten as

$$\Lambda = \mu \exp\left(-\frac{2\pi}{b\alpha(\mu)}\right). \tag{16.26}$$

This means that the combination on the right-hand side is μ independent. Sometimes people say it is RGI (renormalization group invariant). In fact, this is true only in LLA. The exact RGI formula reads

$$\mu \exp\left[-\int^{\alpha(\mu)} \frac{d\alpha}{\beta(\alpha)}\right] = \text{RGI}\,. \tag{16.27}$$

This important statement follows from (14.11).

The parameter Λ has mass dimension 1 and is usually referred to as the *dynamical scale.* The existence of the dynamical scale is not seen in the classical Lagrangian. It appears at the quantum level, due to the so-called *scale anomaly,* see Sec. 15.3, Eq. (15.7).

16.2.1 *A few words on scale transformation*

Now, let us return to the beginning of Sec. 16.1 and discuss the relation between the normalization point μ and the scaling parameter λ.

Let us ask ourselves what happens if we change the scale of distances and momenta? Say, we could measure lengths in terms of centimeters; then we start measuring in terms of meters, $x \to \lambda^{-1}x$, and $d^4x \to \lambda^{-4}\, d^4x$. Then, accordingly, all momenta $p \to \lambda p$. Free fields scale according to their "normal" dimensions, e.g. for $D = 4$ the mass dimension of the scalar fields ϕ is unity, therefore $\phi \to \lambda\phi$. The action is dimensionless. When we speak of (mass) dimensions of free fields we reason as follows. Consider, for instance, the fermion part of the Lagrangian,

$$\Delta\mathcal{L}_0 = \bar\Psi\left(iD_\mu\gamma^\mu\Psi\right) - m_0\bar\Psi\Psi\,, \qquad S = \int d^4x\Delta\mathcal{L}_0 \tag{16.28}$$

Let us change the length scale as indicated above. Given that the action is dimensionless and examining the first term we conclude that under this change

$$\Psi \to \lambda^{3/2}\Psi\,, \tag{16.29}$$

implying

$$m \to \lambda m\,. \tag{16.30}$$

Thus the canonic ("normal") dimension of the mass term is 1, of the Ψ field $\frac{3}{2}$, and so on.

Now, let us return to Eqs. (16.17) and (16.14). Why is γ referred to as the *anomalous dimension*? Quantum corrections bring in a "hidden" dependence on the scale. Indeed, if we fix the UV data, and appropriately scale μ

$$\mu \to \lambda\mu\,, \tag{16.31}$$

we observe from (16.17) that there is an additional scale dependence coming from $\alpha_S(\mu)$.

These anomalous dimensions are in one-to-one correspondence with the scale anomaly in the trace of the energy-momentum tensor of which we will speak later.

Chapter 17

Lecture 17. More on χSB. Effective Lagrangians

Massless fermions and patterns of the chiral symmetry breaking. Chiral Lagrangians at low energies. Gell-Mann and Lévy. The most popular Principle Chiral Model. Baryons in the Skyrme model. Baryon current in the Skyrme model. Non-standard patterns of χSB.

17.1 Spontaneous breaking of chiral symmetry

To begin with, I recall some basic facts regarding Quantum Chromodynamics (QCD).

At low energies QCD can be described as Yang-Mills theory with two or three light Dirac fermions q in the fundamental representation of SU(N). In our actual world the number of colors is $N = 3$. To a good approximation we can consider the light quarks to be massless. Then the QCD Lagrangian

$$\begin{aligned}\mathcal{L} &= -\frac{1}{4} G^{a}_{\mu\nu} G^{\mu\nu\, a} + \sum_{f=1}^{n_f} \bar{q}_f \gamma^{\mu}\, iD_{\mu} q^f \\ &= -\frac{1}{4} G^{a}_{\mu\nu} G^{\mu\nu\, a} + \sum_{f=1}^{n_f} \left[(\bar{q}_L)_f\, \gamma^{\mu}\, iD_{\mu}\, (q_L)^f + (\bar{q}_R)_f\, \gamma^{\mu}\, iD_{\mu}\, (q_R)^f \right]\end{aligned} \tag{17.1}$$

where $G^{a}_{\mu\nu}$ is the gluon field strength tensor, and n_f is the number of the massless flavors (two or three in the actual world). Equation (17.1) is written in the limit of massless quarks. It is important that in this limit the left- and right-handed components of the Dirac spinors representing quarks completely decouple from each other. Hence, the global symmetries which can be seen in the Lagrangian are independent of each other. They correspond to independent flavor rotations in the q_L and q_R sectors.

The global symmetry of the Lagrangian (17.1) is

$$\mathrm{SU}(n_f)_\mathrm{L} \times \mathrm{SU}(n_f)_\mathrm{R} \times \mathrm{U}(1)_V \,. \tag{17.2}$$

The $\mathrm{SU}(n_f)$ symmetries above are *chiral.* The Greek word *chirality* can be translated as *handedness.*

The vectorial U(1), the last factor in Eq. (17.2), is responsible for the baryon number conservation. The *baryon current* is

$$J_\mu^B = \frac{1}{N}\sum_{f=1}^{n_f} \bar{q}_f \gamma_\mu q^f \,. \tag{17.3}$$

I remind that N is the number of colors. An analogous axial current

$$J_\mu^5 = \frac{1}{N}\sum_{f=1}^{n_f} \bar{q}_f \gamma_\mu \gamma^5 q^f \tag{17.4}$$

is anomalous (we will return to this issue later, in Chapter 18) at the quantum level and, therefore is not conserved. The axial-vector U(1) symmetry is nonexistent. The current (17.3) is the sum of the left- and right-handed currents, while (17.4) is the difference.

The fermion term in (17.1) exhibits invariance of the QCD Lagrangian under independent SU(n) rotations of the left- and right-handed quarks, $q_{\mathrm{L,R}} = (1 \mp \gamma_5) q_{\mathrm{L,R}}/2$,

$$q_\mathrm{L}^f \to L_g^f \, q_\mathrm{L}^g \,, \qquad q_\mathrm{R}^{\bar{f}} \to R_{\bar{g}}^{\bar{f}} \, q_\mathrm{R}^{\bar{g}} \,, \tag{17.5}$$

where L and R are the $\mathrm{SU}(n_f)_{\mathrm{L,R}}$ matrices. To emphasize their independence I use barred and unbarred flavor right- and left-handed indices, respectively.[1]

As we will see later, the chiral $\mathrm{SU}(n_f)_\mathrm{L} \times \mathrm{SU}(n_f)_\mathrm{R}$ symmetry is spontaneously broken down to the diagonal $\mathrm{SU}(n_f)_V$,

$$\mathrm{SU}(n_f)_\mathrm{L} \times \mathrm{SU}(n_f)_\mathrm{R} \to \mathrm{SU}(n_f)_V. \tag{17.6}$$

Only the vectorial $\mathrm{SU}(n_f)_V$ is realized linearly in QCD and is seen in the spectrum. The axial part $\mathrm{SU}(n_f)_A$ is spontaneously broken (we understand, of course, that it would be more accurate to say that $\mathrm{SU}(n_f)_A$ is realized nonlinearly; spontaneous breaking is a jargon for a nonlinear realization). The above spontaneous breaking implies the existence of $n_f^2 - 1$ Goldstone bosons, massless pions. Below, I will focus on the case of two massless flavors, $n_f = 2$.

[1] As I explained on page 49, in the spinor representation, the left-handed fermions carry undotted indices, $q_L \to \xi_\alpha$ while the right-handed spinors carry dotted indices, $q_R \to \bar{\eta}^{\dot{\alpha}}$.

17.2 Effective Lagrangians

In this case there are three pion fields $\pi^a(x)$ ($a = 1, 2, 3$). The pion dynamics is concisely described by an SU(2) matrix field $U(x)$,

$$U(x) = \exp\left\{\frac{i}{F_\pi}\,\tau^a\,\pi^a(x)\right\}, \qquad U \in \mathrm{SU}(2)\,, \tag{17.7}$$

where τ^a are the Pauli matrices and

$$F_\pi \approx 93 \text{ MeV}$$

is a so-called pion constant. Another popular parametrization of the matrix U in (17.9) is [2]

$$U(x) = A + i\sum_a \tau^a\,B^a\,.$$

For yet another frequently used representation see Problem 26 on page 270.

Under an SU(2) transformation by unitary matrices (L and R), U transforms as

$$U \to L\,U\,R^\dagger\,. \tag{17.8}$$

The Lagrangian (usually referred to as the *chiral* Lagrangian) must be invariant under both transformations, while the vacuum state must respect only the diagonal combination $L = R$. The Lagrangian must be expandable in powers of derivatives. The lowest-order term has two derivatives and can be written as

$$\mathcal{L}^{(2)} = \frac{F_\pi^2}{4}\,\mathrm{Tr}\left(\partial_\mu U\,\partial^\mu U^\dagger\right)\,. \tag{17.9}$$

It dates back to the old work of Gell-Mann and Lévy [1] who invented the notion of the sigma model. The coordinate space (R^4 in the case on hand) is called the base manifold, while the space on which the fields "live" ($SU(2) \sim S_3$ in the case on hand) is called the target space. Thus, $U(x)$ maps R^4 onto S_3.

The invariance of (17.9) under the global transformations (17.8) is obvious.

Effective Lagrangians of this type represent the expansion of interactions of gapless excitations (massless Goldstones) in the number of derivatives. Equation (17.9) has two derivatives.

---

[2] Please, find a relation between $\vec{\pi}$ and A, $\vec{B}$ and then find an ensuing constraint on A, $\vec{B}$.

In the fifty years that elapsed since their emergence in particle physics, various generalizations of (17.9) spread all over theoretical and mathematical physics, including condensed matter and string theory. For instance, if the target space of the sigma model on hand is S_2, we get the Heisenberg model of (anti)-ferromagnetism (see Sec. 13.2). If the target space is $SU(N)/SU(N-1)\times U(1)$, we get the so-called $CP(N-1)$ models and so on.

17.2.1 *Baryons. The Skyrme model*

The utmost importance in particle physics is the pion-nucleon interactions. Gell-Mann and Lévy suggested [1] to develop the pion-nucleon low-energy theory based on phenomenology. Since then a comprehensive amount of literature on this issue has been developed. I will briefly outline an alternative approach, the so-called Skyrmions in the role of baryons. Skyrmions are topologically stable solitons of a special type supported by some sigma models.

At first, let us introduce a new term to supplement (17.9) the role of which will become clear shortly. In the fourth order in derivatives one can write in the chiral Lagrangian quite many terms invariant under (17.8). As an example, I will present here one such term with four derivatives,

$$\mathcal{L}^{(4)} = \frac{1}{32e^2}\,\mathrm{Tr}\Big[\,(\partial_\mu U)\,U^\dagger\,,\,(\partial_\nu U)\,U^\dagger\Big]^2. \tag{17.10}$$

This operator goes under the name of the *Skyrme term* because Tony Hilton Royle Skyrme invented it. It is singled out because it is the only ∂^4 term that is of the second order in the time derivative. The constant e^2 in Eq. (17.9) is a dimensionless parameter, $e\sim 4.8$. In what follows we will need to know the N dependence of e (here N is the number of colors). I will argue that $e\sim 1/\sqrt{N}$.

Any constant (x-independent) matrix U represents the lowest-energy state, the vacuum of the theory. Each matrix U represents a point in the space of vacua, which is usually referred to as the *vacuum manifold* (in the case on hand, it is the same as the target space). Performing a generic chiral transformation we move from one point of the vacuum manifold to another. However, some chiral transformations, applied to a given vacuum, leave it intact. It is not difficult to understand that all vacua of the theory are invariant under the diagonal $\mathrm{SU}(n_f)_V$ of the chiral $\mathrm{SU}(n_f)_\mathrm{L}\times\mathrm{SU}(n_f)_\mathrm{R}$ group. The easiest way to see this is to consider the vacuum $U = 1$. It is obviously invariant under (17.8) provided $R = L$. Thus, the vacuum manifold is the coset

$$\{\mathrm{SU}(n_f)_\mathrm{L} \times \mathrm{SU}(n_f)_\mathrm{R}\}/\mathrm{SU}(n_f)_V\,. \tag{17.11}$$

The chiral Lagrangian (17.9) describes a $\{\mathrm{SU}(n_f)_\mathrm{L} \times \mathrm{SU}(n_f)_\mathrm{R}\}/\mathrm{SU}(n_f)_V$ *sigma model* in the lowest (second) order in derivatives.

The chiral transformations (17.8) generate flavor nonsinglet currents. As we know from the microscopic theory, there is another conserved current, the baryon current (17.3). What happens to the baryon current in chiral theory (17.9)?

Needless to say, the baryon charge identically vanishes in the meson sector. Thus, if there is a "projection" of the baryon current (17.3) in chiral theory, its expression in terms of U must obey the following property: it must identically vanish for all fields presenting small oscillations of U around $U = 1$.

Such a conserved current does indeed exist,

$$J^{B\,\mu} = -\frac{\varepsilon^{\mu\nu\alpha\beta}}{24\pi^2}\,\mathrm{Tr}\left(U^\dagger\partial_\nu U\right)\left(U^\dagger\partial_\alpha U\right)\left(U^\dagger\partial_\beta U\right)\,, \tag{17.12}$$

and the baryon charge B takes the form

$$B = -\frac{\varepsilon^{ijk}}{24\pi^2}\int d^3x\ \mathrm{Tr}\left(U^\dagger\partial_i U\right)\left(U^\dagger\partial_j U\right)\left(U^\dagger\partial_k U\right)\,. \tag{17.13}$$

I leave it as an exercise to prove that the current (17.12) is conserved "topologically" (i.e. one does not need to use equations of motion in the proof) and that $B \equiv 0$, order by order in the expansion of (17.7) in the fields π assuming $|\pi| \ll 1$ and $\pi(x) \to 0$ at $|\vec{x}| \to \infty$.

Now, let us consider $\vec{x}$ dependent but static field configuration [3]

$$U_\mathrm{Sk}(\vec{x}) = \exp\left(i\,F(r)\,\frac{\tau^j\,x_j}{r}\right)\,, \qquad r = |\vec{x}|\,, \tag{17.14}$$

where a dimensionless function $F(r)$ parametrizes the Skyrmion profile. Moreover, τ^j are three Pauli matrices. We see that the $SU(2)$ flavor is entangled with the coordinate space R_3. This is a typical hedgehog according to the Polyakov terminology. For the function (17.14) to be regular at the origin and tend to a constant at the spatial infinity

[3] If the number of massless flavors is three or larger, the so-called Wess-Zumino-Novikov-Witten (WZNW) term must be added to the chiral Lagrangian (17.9). In this course we will not discuss it. For two flavors the Wess-Zumino-Novikov-Witten term identically vanishes.

(this requirement is necessary and sufficient to guarantee finiteness of energy) we must impose the conditions

$$F(r) = \pi \times \text{integer at } r = 0 \text{ and } F(r) = \pi \times \text{integer at } r \to \infty\,. \tag{17.15}$$

Important Statement: if the two integers mentioned in (17.15) do *not* coincide with each other then the field configuration (17.14) cannot be continuously deformed to a constant matrix U. This means that it is topologically stable. This field configuration is not reducible to a finite number of pions.

Let us check its baryon charge. Substituting (17.14) in the definition of the baryon charge (17.13), after a straightforward algebra, we reduce the integrand to a full derivative, and find

$$B = -\frac{1}{\pi}\left[F(r) - \frac{1}{2}\sin 2F(r)\right]_0^\infty\,. \tag{17.16}$$

To obtain the baryon charge, one can set

$$F(0) = \pi\,, \quad F(\infty) = 0\,. \tag{17.17}$$

If we can get such a solution from the chiral Lagrangian-derived equations, we will get nucleons in the same model!

If we just have the two-derivative chiral Lagrangian then the solution is singular, its size shrinks to zero. However, adding the Skyrme term (17.10) stabilizes the size of the Skyrmion. It is convenient to measure r in its natural units. Therefore, we will introduce a dimensionless "radius" ρ instead of r,

$$\rho = eF_\pi\, r\,. \tag{17.18}$$

Not only is ρ dimensionless, it is also N independent. Indeed, $F_\pi \sim \sqrt{N}$ while $e \sim 1/\sqrt{N}$. Due to this fact, the Skyrmion size is N independent too – a feature which is most welcome because (as we know from the quark picture) the baryon size does not scale with N at large N. The profile function $F(\rho)$ was calculated numerically and is shown in Fig. 17.1. To this end one substitutes (17.14) in (17.9) and (17.10) and then minimizes the energy functional over $F(\rho)$ under the boundary conditions (17.17). In this way, one obtains the Skyrmion mass,

$$M_{\rm sk} \approx 1.23\left(6\pi^2\frac{F_\pi}{e}\right). \tag{17.19}$$

The Skyrmion mass scales as N at large N because so does the expression in parentheses in (17.19). This is in full accord with the

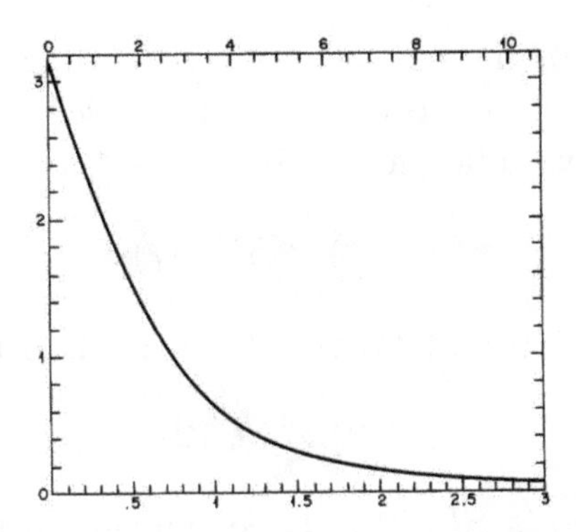

Fig. 17.1 The Skyrme profile function *versus* ρ.

quark-based expectation. Even at $N = 3$ it is reasonably close to its experimental value.

A few words are in order to answer the question on why is the above quasiclassical description of the baryons justified and should work at large N. The quasiclassical description works well at weak coupling. The loop expansion parameter in the theory under consideration is

$$\frac{1}{(4\pi F_\pi)^2} \sim \frac{1}{N}.$$

At large N, it is arbitrarily small. At $N = 3$, one can expect only a qualitative agreement at best.

17.3 χSB versus color and flavor

It is worth noting that the pattern of the chiral symmetry breaking (χSB) depends – although indirectly – on the number of colors and on the representation of the fermion fields. Above, we have considered $SU(3)_{\rm c}$ and the fermion fields in the fundamental representation of the gauge group.

Now, let us briefly discuss a few examples of other χSB patterns, which can serve as a good theoretical laboratory.

(a) $SU(2)_{\rm c}$, *Dirac fermions in the fundamental representation*

Each Dirac fermion consists of one dotted and one undotted Weyl spinor (i.e. one left-handed and another right-handed), which can be written, e.g. as

$$\Psi = \begin{pmatrix} \chi_\alpha \\ \bar\eta^{\dot\alpha} \end{pmatrix} \tag{17.20}$$

where color and flavor indices are omitted. For $SU(3)_c$ two spinors, χ and η form different color representations, since the first is triplet while the second antitriplet. Therefore the flavor symmetry of the model is[4]

$$SU(n_f) \times SU(n_f)\,. \tag{17.21}$$

If we take $n_f = 2$ for simplicity, the flavor symmetry of QCD is

$$SU(2) \times SU(2)\,. \tag{17.22}$$

If the gauge group is $SU(2)_c$, the situation changes, since the color doublet can be converted into the antidoublet by virtue of the Levy-Civita tensor,

$$\varepsilon^{ij}\eta_{j\,\alpha}\,, \quad i, j \text{ are } SU(2)_c \text{ color indices.} \tag{17.23}$$

Therefore, the flavor symmetry in this case is $SU(2n_f)$ rather than $SU(n_f) \times SU(n_f)$. For $n_f = 2$, the pattern of χSB is as follows:

$$SU(4) \to SO(5)\,. \tag{17.24}$$

If for any number of colors we take n_f Dirac spinors in the adjoint representation, the pattern of χSB is as follows:

$$SU(2n_f) \to SO(2n_f)\,. \tag{17.25}$$

This is because the adjoint representation is real: its complex conjugate coincides with itself.

More details can be found in [2].

I conclude this subsection with the following intriguing question: can one obtain baryon as a quasiclassical soliton in QCD with a single quark (i.e. one Dirac field in the fundamental representation)?

QCD with one quark obviously does have baryons. At the same time, there is no chiral symmetry and no effective chiral Lagrangian. It seems that in the absence of a Skyrme-like model the solitonic description of baryons is impossible. This was a general belief for over 30 years. However, Zohar Komargodski argued [3] that the solitonic description of baryons can be built in this case as well, although it is more complicated than the good old Skyrmions. Komargodski's construction goes well beyond this lecture course.

The interested reader is referred to [3].

[4]For the time being I omit $U(1)$. Later we will discuss $U(1)$ symmetries separately.

17.4 The number of Goldstone particles (gapless excitations)

From the previous semester we know that the number of Goldstone particles ν (let us generically call such particles π; in CM physics they are referred to as gapless excitations) is related to the number of broken generators. This relation is different, however, for relativistic and non-relativistic field theories,

$$\nu_\pi = \begin{cases} \nu_G - \nu_H & \text{Relativistic QFT} \\ \frac{1}{2}(\nu_G - \nu_H) & \text{Non-relativistic QFT,} \end{cases} \tag{17.26}$$

where ν_G and ν_H are the numbers of generators in the groups G and H respectively. The additional factor $\frac{1}{2}$ in the second line appears when we quantize the theory.

Indeed, at the classical level the number of the gapless excitations is $\nu_G - \nu_H$. Let us see what happens with a single gapless excitation ϕ after quantization. In the relativistic case the time derivative enters quadratically,

$$\mathcal{L}_{\rm r} = \frac{1}{2}\left[(\dot\phi)^2 - (\vec\nabla\phi)^2\right] \tag{17.27}$$

where $\phi(t,\vec x)$ is a real field. Canonic quantization implies that

$$\pi_\phi(t,\vec x) = \dot\phi(t,\vec x)\,, \tag{17.28}$$

where π_ϕ is a canonic momentum. The pair $\{\pi_\phi\,,\,\phi\}$, with independent π_ϕ and ϕ, describes exactly one massless quantum field.

Now, in the non-relativistic case, the time derivative enters linearly, for instance

$$\mathcal{L}_{\rm nr} = \phi(\dot\phi) - (\vec\nabla\phi)^2\,. \tag{17.29}$$

Since $\phi(\dot\phi) = \frac{1}{2}(\dot{\phi^2})$, the time derivative drops off from the action. Roughly speaking, we deal with a half of a quantum degree of freedom, which is invalid in QFT.

Assume, however, that we have two classical gapless modes in a non-relativistic theory, ϕ_1 and ϕ_2. Then we can always build a complex field

$$\varphi = \frac{1}{\sqrt2}\left(\phi_1 + i\phi_2\right)\,. \tag{17.30}$$

The Lagrangian then takes the form

$$\mathcal{L}_{\rm nr} = \varphi^\dagger(\dot\varphi) - (\vec\nabla\varphi^\dagger)(\vec\nabla\varphi)\,, \tag{17.31}$$

Fig. 17.2 Murray Gell-Mann in 2012.

and $\pi_\varphi = \varphi^\dagger$ and φ form a canonic pair (please, note, a single canonic pair, rather than two pairs in the relativistic case) and represents one massless quantum.

This is the reason behind the occurrence of $\frac{1}{2}$ in the second line in Eq. (17.26).

Appendix 17.1: On Maurice Lévy and Tony Skyrme

Maurice Lévy (b. 1922) was born in Tlemcen, Algeria, to a long-established Jewish family. In 1837 his great-grandfather, Maklouf Lévy, born in 1812, had served as a translator in the negotiations between General Bugeaud and Emir Abd-el-Kader for the surrender of Tlemcen. As a reward, he had received French nationality for himself and his family, 33 years before the Cremieux decree, which granted French nationality to the Jews of Algeria.

***

Fig. 17.3 Maurice Lévy in the 1970s.

"I don't know how many people are aware of the impish side of Maurice's nature. He was at the blackboard once when he recalled the time in 1960 when he and Murray Gell-Mann were writing a paper which introduced the so-called sigma-model; it later became famous. Murray was scheduled to give a talk on their work at the Collége de France the next day. While going over their work and writing an equation on the blackboard, Maurice referred to two terms which, had he been speaking English, he would have said "are going to cancel." But he was speaking French and under the influence of some impishness he said "ils vont se canceler." Murray took notice and when he came to the same point at the seminar, he used that phrase. The crowd tittered and Murray, who was justly proud of his command of French, soon discovered that the correct expression is "ils vont s'annuler." He had been led up the garden path by Maurice. "Canceler" is not a French verb, it

Fig. 17.4 Tony Hilton Royle Skyrme (1922-1987). Manhattan project (1944-45), then Oxford and University of Birmingham, MIT (1948-49), Princeton (1949-50), Harwell (1950-62), University of Malaya (1962-64), University of Birmingham. He invented Skyrmions ahead of their time.

just sounds French (it's not even an English verb). Years later, when Gell-Mann was planning a bird-watching safari in Africa, he asked Lévy if he could recommend a reliable outfit. Maurice mentioned, in all innocence, a couple he had heard of. It turned out that the one Murray chose was a complete disaster. After that, according to Maurice, Gell-Mann "never fully trusted him."

References

[1] Murray Gell-Mann and M. Levy, *The axial vector current in beta decay*, Nuovo Cim. **16**, 705 (1960).

[2] Y. I. Kogan, M. A. Shifman and M. I. Vysotsky, *Spontaneous breaking of chiral symmetry for real fermions and $\mathcal{N} = 2$ SUSY Yang-Mills theory*, Sov. J. Nucl. Phys. **42**, 318 (1985).

[3] Z. Komargodski, "Baryons as Quantum Hall Droplets," arXiv:1812.09253 [hep-th].

Chapter 18

Lecture 18. Quantum Anomalies

Classical symmetries and quantum anomalies in four-dimensional Yang-Mills theories. Triangle anomalies in the axial current obtained in the framework of two different regularizations. I will not discuss diangle anomalies in two-dimensional theories (see, however, Problem 27b, page 273). Nor will I derive the scale anomaly. Wilson line and Wilson loop are introduced.

This topic is important, since the phenomenon of anomalies plays a role in a number of subtle aspects of gauge dynamics. It also provides us with one of the few tools available at strong coupling.

Anomalies can appear in a number of reincarnations. "Internal" anomalies kill the theory under consideration; they should be cancelled (e.g. as in the Standard Model, SM, see Chapter 19). "External" anomalies are harmless for the theory *per se* and, moreover, they can be used to obtain certain information which would be impossible by other means. Finally, there is a rare kind of anomaly which could be called "global." It can be traced to Witten [1; 2]. I will mention them below but won't go into details.

A quantum anomaly usually appears under the following circumstances. Assume the theory under consideration has two or more symmetries at the classical level. However, when you start quantizing it you discover that at the quantum level you can maintain only one symmetry and must sacrifice the other one. In some instances, you can choose what symmetry to keep and what to give up. But if one of these two symmetries is the gauge symmetry,[1] there is no choice: you have to keep the gauge symmetry at all costs, because otherwise your

[1] Remember, the gauge symmetry is not a genuine symmetry, but rather a redundancy in theoretical description.

theory becomes inconsistent. "Internal" anomalies are those to which gauge bosons are coupled. Such anomalies cannot be tolerated.

Most quantum anomalies are related to chirality; they are referred to as chiral anomalies, with which we will start.

The chiral anomaly naturally involves fermions.[2] Implications of the chiral anomalies are important. We will discuss the 't Hooft matching condition, one of the few tools applicable to non-Abelian theories at strong coupling. Later we will prove that the chiral symmetry of QCD must be spontaneously broken. As an illustrative example of the usefulness of the proper understanding of the anomalies, we will calculate the $\pi^0 \to \gamma\gamma$ decay rate (see Chapter 20).

18.1 Anomalies in QCD and similar non-Abelian gauge theories

In this lecture we will discuss QCD and analogous non-Abelian gauge theories which are self-consistent, i.e. free of internal anomalies. In particular, dealing with chiral theories we should follow strict rules in constructing the matter sector.

The chiral anomalies in four dimensions are also referred to as triangle, or Adler-Bell-Jackiw anomalies.[3] Let us consider QCD as a showcase. More exactly, we will assume that the theory under consideration has the gauge group SU(N) and contains n_f massless quarks

---

[2] There is another anomaly in gauge theories, the scale anomaly. It occurs even in pure Yang-Mills theory, with no quarks.

[3] In fact, the anomaly to be discussed below, was first "observed" by Jack Steinberger in 1949, who later became a well-known experimentalist, but was a student in 1949. Here is an excerpt from Roman Jackiw paper "Remembering John Bell" (1991): "One time, Jack Steinberger – John's friend and collaborator on a CP formalism – was at the table and asked about current interests. When he described to him the $\pi^0 \to 2\gamma$ puzzle, Jack expressed amazement that theorists should still be pursuing a process that he, as an experimentalist, calculated almost twenty years earlier, finding excellent agreement with experiment, while also noting a discrepancy between results obtained when the pion coupled to nucleons by pseudo-vector or pseudo-scalar interactions. (Pions, nucleons and photons were the only particles [known] in Steinberger's [student time], and it was believed that equivalent results emerge for pseudo-vector and pseudo-scalar pion-nucleon coupling.) There at that table came to us the realization that Steinberger's calculation would be identical to the one performed in the dynamical framework of the σ-model, which was constructed to realize current algebra explicitly. We reasoned that within the σ-model we could satisfy the current algebraic assumptions of Sutherland/Veltman and also obtain good experimental agreement in view of Steinberger's result – thereby resolving $\pi^0 \to 2\gamma$."

(Dirac fields in the fundamental representation). In this section, it will be convenient to write the action in the canonic normalization,

$$S = \int d^4x \left\{ -\frac{1}{4} G^a_{\mu\nu} G^{\mu\nu\, a} + \sum_{f=1}^{n_f} \bar{\psi}_f \, i \not{D} \, \psi^f \right\}. \tag{18.1}$$

We will start from examining the classical symmetries of the above action.

In addition to the scale (implying, in fact, full conformal) invariance of the classical action of which I will speak later, (18.1) has the following symmetry

$$\mathrm{U}(1)_V \times \mathrm{U}(1)_A \times \mathrm{SU}(n_f)_L \times \mathrm{SU}(n_f)_R \tag{18.2}$$

acting in the matter sector. The vector U(1) corresponds to the baryon number conservation, with the current

$$j^B_\mu = \frac{1}{3} \bar{\psi}_f \, \gamma_\mu \, \psi^f . \tag{18.3}$$

The axial U(1) corresponds to the overall chiral phase rotation[4]

$$\psi^f_L \to e^{i\alpha} \psi^f_L , \qquad \psi^f_R \to e^{-i\alpha} \psi^f_R , \qquad \psi_{L,R} = \frac{1}{2}(1 \mp \gamma^5)\psi . \tag{18.4}$$

The axial current generated by (18.4) is

$$j^\mu_A = \bar{\psi}_f \, \gamma^\mu \, \gamma^5 \, \psi^f . \tag{18.5}$$

At the quantum level (i.e. including loops with a regularization) the fate of the above symmetries is different. The vector U(1) invariance generated by (18.3) stays a valid anomaly-free symmetry at the quantum level. The axial current is anomalous.

18.2 Chiral anomaly in the singlet axial current

Differentiating (18.5) naively we get $\partial_\mu j^\mu_A = \bar{\psi}_f \overleftarrow{\not{D}} \, \gamma^5 \psi^f - \bar{\psi}_f \gamma^5 \not{D} \psi^f = 0$ by virtue of the equation of motion $\not{D}\, \psi^f = 0$. The classical axial current conservation will not hold upon switching on a gauge-symmetry respecting regularization. To make the calculation of the anomaly reliable we must exploit only the Green functions at short distances. This means that we must focus directly on $\partial_\mu j^\mu_A$ exploiting one of appro-

[4]See footnote on page 178.

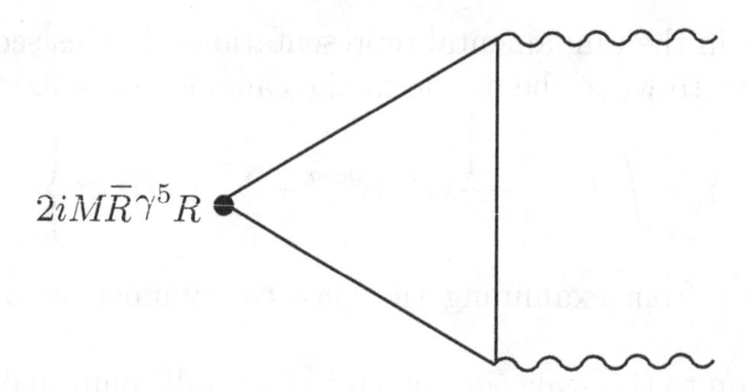

Fig. 18.1 Diagrammatic representation of the triangle anomaly. The solid and wavy lines denote the regulator and gluon fields, respectively.

priate ultraviolet regularizations. The following demonstration will be based on the Schwinger and Pauli–Villars regularizations. [5]

18.2.1 *The Pauli–Villars regularization*

Let us introduce the Pauli–Villars fermion regulators R with mass M_R to be sent to infinity at the very end. Then the regularized singlet axial current takes the form

$$j_\mu^{A,\,R} = \bar{\psi}_f\,\gamma_\mu\gamma^5\,\psi^f + \bar{R}_f\,\gamma_\mu\gamma^5\,R^f\,. \tag{18.6}$$

Since the current is now regularized, its divergence can be calculated according to the equations of motion,

$$\partial^\mu\,j_\mu^{A,\,R} = 2i\,M_R\,\bar{R}_f\gamma^5 R^f\,. \tag{18.7}$$

As was expected, the result contains only the regulator term. Our next task is to project it onto "our" sector of the theory in the limit $M_R \to \infty$. In this limit only the two-gluon operator will survive, as depicted in the triangle diagram of Fig. 18.1.

*****

This diagram can be calculated either by the standard Feynman graph technique, or using the background field, which is quite straight-

[5]The widely used dimensional regularization is awkward and inappropriate in those problems in which γ^5 is involved.

forward in the case on hand,[6]

$$2i\, M_R\, \bar{R}_f \gamma^5 R^f \to 2i\, M_R n_f \mathrm{Tr}_{\mathrm{C,L}} \left(\gamma^5 \, \frac{i}{i\not{D} - M_R} \right)$$

$$\to -2\, M_R n_f \mathrm{Tr}_{\mathrm{C,L}} \left[\gamma^5 \, \frac{1}{(i\mathcal{D})^2 - M_R^2 + \frac{ig}{2}\, G_{\mu\nu}\sigma^{\mu\nu}} \, (i\not{D} + M_R) \right] . \quad (18.8)$$

Here we omitted an extra minus sign which would have been necessary if it were an ordinary fermion loop. Given that the triangle loop in Fig. 18.1 is that of the regulator fields, the extra minus sign must not be inserted. The term $i\not{D}$ in the last brackets can be dropped because of the trace with γ^5. Remembering that $M_R \to \infty$, one can expand the denominator in $G\sigma$. The zeroth-order term in this expansion vanishes for the same reason. The term $O(G\sigma)$ vanishes because of the color trace. The term $O((G\sigma)^2)$ does not vanish, while all higher order terms are suppressed by positive powers of $1/M_R$ and disappear in the limit $M_R \to \infty$. In this way, we arrive at

$$2i\, M_R\, \bar{R}_f \gamma^5 R^f \to \frac{M_R^2\, g^2}{2}\, n_f \mathrm{Tr}_{\mathrm{C}} \left(G_{\mu\nu} G_{\alpha\beta} \right) \mathrm{Tr}_{\mathrm{L}} \left(\gamma^5 \sigma^{\mu\nu} \sigma^{\alpha\beta} \right)$$

$$\times \int \frac{d^4 p}{(2\pi)^4} \, \frac{1}{(p^2 - M_R^2)^3} \,, \quad (18.9)$$

which, in turn, implies that

$$\partial^\mu \, j_\mu^A = n_f \, \frac{g^2}{16\pi^2} \, G^{\alpha\beta\, a} \, \tilde{G}^a_{\alpha\beta} \, . \quad (18.10)$$

The above relation is referred to as the chiral anomaly. It is exact in the sense that two-, three-, and higher-loop corrections do not appear on the right-hand side. Equation (18.10) is in full accord with the result (18.14) to be obtained below in the Schwinger regularization. Characteristic distances saturating the triangle loop in Fig. 18.1 are of the order of $M_R^{-1} \to 0$ at $M_R \to \infty$.

[6] The transition from the first to second line in (18.8) is due to the following equalities:

$$(i\not{D} - M_R)\,(i\not{D} + M_R) = i\not{D}\, i\not{D} - M_R^2 \, ;$$

and

$$\gamma^\mu \gamma^\nu = g^{\mu\nu} + \frac{1}{2}\, [\gamma^\mu \, \gamma^\nu] = g^{\mu\nu} + \sigma^{\mu\nu} \, .$$

18.2.2 *The Schwinger regularization**

In this regularization we ε-split the current,

$$j_\mu^{A,\,R}(x) = \bar{\psi}_f(x+\varepsilon)\,\gamma_\mu\,\gamma^5 \left\{ P\exp\left[\int_{x-\varepsilon}^{x+\varepsilon} i\,A_\rho(y)\,dy^\rho\right]\right\}\psi^f(x-\varepsilon)\,. \tag{18.11}$$

Here the superscript R marks the regularized current while $A_\rho \equiv A_\rho^a\,T^a$. The ε parameter must be set to zero at the very end. The exponent is necessary to ensure gauge invariance of the regularized current $j_\mu^{A,\,R}$ after the split

$$\bar{\psi}_f(x+\varepsilon)\,\psi^f(x-\varepsilon)\,. \tag{18.12}$$

In more detail, we will discuss this exponent at the end of this lecture.

Next, we differentiate over x using the equations of motion above. Expanding in ε and keeping terms $O(\varepsilon)$ we arrive at

$$\partial^\mu j_\mu^{A,\,R} = \bar{\psi}_f(x+\varepsilon)\Big[-ig\not{A}(x+\varepsilon)\,\gamma^5 - \gamma^5\,ig\not{A}(x-\varepsilon)$$
$$+\,ig\,\gamma^\mu\gamma^5\varepsilon^\beta\,G_{\mu\beta}(x)\Big]\,\psi^f(x-\varepsilon)\,. \tag{18.13}$$

The third term in the braces in (18.13) contains the gluon-field strength tensor. This term is due to the differentiation of the exponential factor. The gluon 4-potential A_μ and the field strength tensor $G_{\mu\beta}$ are treated as background fields. For small values of y one can always choose the following representation for the background field:

$$A_\mu(y) = \frac{1}{2}y^\rho\,G_{\rho\mu}(0) + \ldots\,.$$

Now, we contract the quark lines (18.12) to form the quark Green's function $S(x-\varepsilon,\,x+\varepsilon)$ in the background field and then we arrive at (see (18.16))

$$\partial^\mu j_\mu^{A,\,R} = -ig\,n_f\,\mathrm{Tr}_{\mathrm{C,L}}\left\{-2i\,\varepsilon^\rho\,G_{\rho\mu}(x)\,\gamma^\mu\gamma^5\,S(x-\varepsilon,\,x+\varepsilon)\right\}$$
$$= -n_f\,\frac{g^2}{2}\,G_{\rho\mu}(x)^a\,\tilde{G}_{\alpha\phi}(x)^a\,\frac{\varepsilon^\rho\varepsilon^\alpha}{\varepsilon^2}\,\frac{1}{8\pi^2}\,\mathrm{Tr}_{\mathrm{L}}\left(\gamma^\mu\gamma^5\gamma^\phi\gamma^5\right)$$
$$= \frac{n_f\,g^2}{16\pi^2}\left\{G^{\alpha\beta\,a}\,\tilde{G}^a_{\alpha\beta}\right\}_{\mathrm{bckgd}}, \tag{18.14}$$

where

$$\tilde{G}_{\alpha\beta} = \frac{1}{2}\varepsilon_{\alpha\beta\rho\mu}G^{\rho\mu}\,, \tag{18.15}$$

and the subscripts C and L mark the traces over the color and Lorentz indices, respectively. The most crucial point is that the Green's function $S(x-\varepsilon,\, x+\varepsilon)$ is used *only* at very short distances $\sim \varepsilon \to 0$, where it is reliably known in the form of an expansion in the background field. We need only the first nontrivial term in this expansion),

$$S(x,\,y) = \frac{1}{2\pi^2}\frac{\not{r}}{(r^2)^2} - \frac{1}{8\pi^2}\frac{r^\alpha}{r^2}\, g\, \tilde{G}_{\alpha\phi}(x)\, \gamma^\phi \gamma^5 + ..., \qquad r = x - y\,. \tag{18.16}$$

In passing from the second to the third line in Eq. (18.14), averaging over the angular orientations of the four-vector ε is performed.

18.3 The chiral anomaly for generic fermions

What changes occur in the chiral anomaly if instead of the fundamental representation, we consider fermions in some other representation R? The answer to this question is simple. If we inspect derivations in Secs. 18.2.2 and 18.2.1, we will observe that the result for the anomalous divergence of the axial current is proportional to $\mathrm{Tr} T^a T^b$. For the fundamental representation in SU(N)

$$\mathrm{Tr} T^a T^b = \frac{1}{2}\delta^{ab}\,.$$

In the general case

$$\mathrm{Tr} T^a T^b = T(R)\delta^{ab}\,,$$

where $T(R)$ is (one half) of the Dynkin index for the given representation. Thus, if we have n_f massless Dirac fermions in the representation R, then Eq. (18.10) must be replaced by the following formula:

$$\partial^\mu \left(\bar{\psi}_f \gamma_\mu \gamma^5 \psi_f\right) = n_f\, \frac{T(R)\, g^2}{8\pi^2}\; G^{\alpha\beta\, a}\, \tilde{G}^a_{\alpha\beta}\,. \tag{18.17}$$

For instance, for the adjoint representation in SU(N) one has $T(\mathrm{adj}) = N$. Note that for the real representations, such as the adjoint, one can consider not only Dirac fermions, but Majorana fermions as well. Each Majorana fermion counts as $n_f = \frac{1}{2}$. The same is true with regards to the Weyl fermions with which one has to deal in chiral Yang-Mills theories.

Example: If ψ is the Majorana field in the adjoint representation, the divergence of the axial current is

$$n_f\, \frac{N\, g^2}{16\pi^2}\; G^{\alpha\beta\, a}\, \tilde{G}^a_{\alpha\beta}\,.$$

18.4 Wilson line and Wilson loop

Let us return to Eq. (18.11) and have a closer look at the exponent inserted between ϵ-splitted $\bar{\psi}$ and ψ. In QED, Julian Schwinder defined [3] a nonlocal operator

$$U(x_2, x_1) \equiv \exp\left[\int_{x_1}^{x_2} i\, A_\mu(x)\, dx^\mu\right]. \tag{18.18}$$

Generally speaking, $U(x_2, x_1)$ depends on the contour C connecting the initial and final points; therefore, it would be more accurate to write $U_C(x_2, x_1)$.

Using the QED gauge transformation law, it is easy to see that [7]

$$\exp\left(i\int_{x_1}^{x_2} A_\mu(x) dx^\mu\right)_{\rm gt} = \exp\left[i\int_{x_1}^{x_2}\Big(A_\mu(x) + \partial_\mu\alpha(x)\Big)\, dx^\mu\right]$$
$$= e^{\alpha(x_2)}\, \exp\left(i\int_{x_1}^{x_2} A_\mu(x) dx^\mu\right) e^{-\alpha(x_1)} \tag{18.19}$$

or

$$U(x_2, x_1)_{\rm gt} = e^{\alpha(x_2)}\, U(x_2, x_1) e^{-\alpha(x_1)}\,. \tag{18.20}$$

The operator (18.18) could have been called *Schwinger line* but it is not. Note that while $U(x_2, x_1)$ is generally dependent on the line geometry, its gauge transformation depends only on α at the line endpoints.

Equation (18.20) explains why (18.11) is gauge invariant.

Wilson generalized (18.18) to non-Abelian gauge theories,

$$U(x_f, x_i) \equiv P\, \exp\left[\int_{x_i}^{x_f} i\, A_\mu(x)\, dx^\mu\right]. \tag{18.21}$$

Currently it is referred to as the Wilson line. Unlike the Abelian case, the field A_μ is an x-dependent matrix, $A_\mu(x) = A_\mu^a(x) T^a$, and the matrices $A_\mu(x)$ and $A_\mu(x')$ generally do not commute if $x \neq x'$. Therefore, we have to insert the symbol of path ordering P in (18.21), analogous to the time ordering T. The closer is x to x_i, the more to the right $A_\mu(x)$ stands in the path-ordered product. We want to prove that the gauge transformation law of the Wilson line is

$$P\, \exp\left[\int_{x_i}^{x_f} i\, A_\mu(x)\, dx^\mu\right]_{\rm gt} = U(x_f) P\, \exp\left[\int_{x_i}^{x_f} i\, A_\mu(x)\, dx^\mu\right] U^\dagger(x_i) \tag{18.22}$$

[7] See Lecture 1, Eq. (1.8), which also can be written as $A_\mu \to U\, A_\mu\, U^\dagger - i\,(\partial_\mu U)\, U^\dagger$ where $U = \exp(i\alpha)$.

where $U(x)$ is the gauge transformation matrix from the gauge group G (which is $SU(N)$ in this lecture course). Equation (18.22) is to be compared with Eq. (18.20).

To prove (18.22) let us first consider the case of $x_2 \to x_1$ where $x_f \equiv x_2$ and $x_i \equiv x_1$. For very close endpoints we can always consider the line to be straight and expand the exponent to the first order in $\epsilon \equiv x_2 - x_1$,

$$U(x_2, x_1) = 1 + iA_\mu(x_1)\epsilon^\mu + ... \tag{18.23}$$

where the dots stand for higher orders in ϵ which can be omitted. Using the gauge transformation of the non-Abelian four-potential from Lecture 1, we obtain

$$U(x_2, x_1)_{\text{gt}} = 1 + iU(x_1)A_\mu(x_1)U^\dagger(x_1)\epsilon^\mu + (\partial_\mu U(x_1))\,U^\dagger(x_1)\epsilon^\mu + ... \tag{18.24}$$

This obviously coincides with (18.22),

$$U(x_2, x_1)_{\text{gt}} = U(x_2)\,U(x_2, x_1)\,U^\dagger(x_1)\,, \tag{18.25}$$

if we expand $U(x_2)$ up to the first order in ϵ,

$$U(x_2) = U(x_1) + \partial_\mu\,U(x_1)\epsilon^\mu + ... \tag{18.26}$$

Next we split the line connecting x_i and x_f into n very close points, $x_i = x_1$, x_2, x_3, ..., x_f keeping in mind to tend $n \to \infty$. Then $U(x_f, x_i)$ from (18.21) can be represented as

$$U(x_f, x_i) = U(x_f, x_{f-1}) \times ... \times U(x_3, x_2) \times U(x_2, x_i) \tag{18.27}$$

and

$$\begin{aligned} U(x_f, x_i)_{\text{gt}} &= \Big(U(x_f)\,U(x_f, x_{f-1})U^\dagger(x_{f-1})\Big) \times ... \\ &\times \Big(U(x_3)\,U(x_3, x_2) \times U^\dagger(x_2)\Big) \times \Big(U(x_2)U(x_2, x_i)U^\dagger(x_i)\Big) \\ &= U(x_f)\,U(x_f, x_i)U^\dagger(x_i)\,, \end{aligned} \tag{18.28}$$

quod erat demonstrandum.

One last remark is in order here. Assume that instead of a line with two endpoints we take a closed contour C, i.e. deform the line in such a way that its two endpoints coincide,

$$U(C) \equiv \text{Tr}\; P\; \exp\left[\oint i\,A_\mu(x)\,dx^\mu\right]. \tag{18.29}$$

Then, obviously, Eq. (18.28) implies that $U(C)$ is gauge invariant. The object in Eq. (18.29) is referred to as the *Wilson contour operator*, or *Wilson loop.*

References

[1] E. Witten, *An SU(2) Anomaly,* Phys. Lett. B **117**, 324 (1982).

[2] G. W. Moore and P. C. Nelson, *The etiology of σ model anomalies*, Commun. Math. Phys. **100**, 83 (1985); J. Chen, X. Cui, M. Shifman and A. Vainshtein, *Anomalies of minimal $\mathcal{N} = (0,1)$ and $\mathcal{N} = (0,2)$ sigma models on homogeneous spaces*, J. Phys. A **50**, no. 2, 025401 (2017).

[3] J. S. Schwinger, Phys. Rev. **82**, 664 (1951).

Chapter 19

Lecture 19. Quantum Anomalies (*Continued*)

> *Internal chiral anomalies in gauge field theories are fatal. Cancellation of internal anomalies in the Standard Model. A few words on Witten's global anomaly in SU(2) Yang-Mills theory.*

19.1 Cancellation of internal anomalies in Standard Model (SM)

In QCD the gauge bosons are gluons. They are coupled to quarks only through γ^μ matrix, not through $\gamma^\mu\gamma^5$. Such theories are called vector-like. It is obvious that because of the absence of the $\gamma^\mu\gamma^5$ coupling, triangle graphs with three gluons in the vertices (i.e. ggg graphs) have no anomalies.

19.2 Internal anomalies in SM. First generation

This is not the case in the Standard Model of electroweak interactions (SM). This model is chiral, i.e. left-handed fermions are coupled to the gauge bosons in a way different from the right-handed coupling. For instance, only the left-handed fermions are coupled to the W bosons. Thus, the anomalous triangle graphs (such as the graph depicted in Fig. 19.1) with one $\gamma^\mu\gamma^5$ vertex and two γ^μ's, with three gauge bosons in the vertices, are present. The only way to save the theory is to cancel individual anomalous graphs in the total sum. Only the first generation of fermion matter will be considered. The second and third generations can be treated in absolutely the same way.

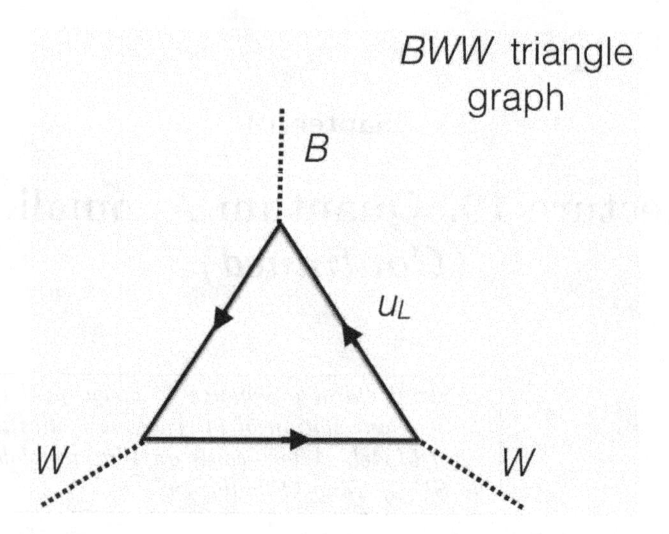

Fig. 19.1 BWW anomalous triangle. Since W's interact only with the left-handed fermions, the fermions propagating in this triangle are: u_L, d_L, ν_L, e_L.

In addition to gluons, in SM there are the following gauge bosons:

$$W^{\pm}\,,\quad W^3\,,\quad B\,, \tag{19.1}$$

where B is the gauge field coupled to weak hypercharge. In fact, due to the spontaneous $SU(2)$ symmetry breaking, the physical gauge bosons are Z and γ, obtained as linear combinations of W^3 and B. For cancellation of anomalies one can check either the first pair of the gauge bosons (i.e. W^3 and B) , or the second (Z and γ). It does not matter.[1] We will examine $W^{\pm}\,,\quad W^3\,,\quad B$.

The potentially anomalous triangle graphs are[2]

$$Bgg,\; W^3W^+W^-,\; W^3W^3W^3,\; BBB,\; BW^3W^3,\; BW^+W^-. \tag{19.2}$$

The BWW example (with u_L in the triangle loop) is presented in Fig. 19.1. We will consider all these triangles one by one. The weak hypercharges of quarks and leptons are determined by the following rule: the weak hypercharge is twice the average electric charge in the

[1]Think and try to figure out why it does not matter!

[2]The triangles WBB and Wgg vanish for trivial reasons: indeed, W fields belong to the adjoint representation of the $SU(2)_{\rm weak}$ group while the B and g fields are singlets with respect to this group. This means that the above two triangles are proportional to Tr T^a =0. Here T^a are the $SU(2)$ generators.

given weak multiplet,[3]

$$\begin{pmatrix} u_L \\ d_L \\ \frac{1}{3} \end{pmatrix}, \quad \begin{pmatrix} u_R \\ \\ \frac{4}{3} \end{pmatrix}, \quad \begin{pmatrix} d_R \\ \\ -\frac{2}{3} \end{pmatrix}, \quad \begin{pmatrix} \nu_L \\ e_L \\ -1 \end{pmatrix}, \quad \begin{pmatrix} \nu_R \\ \\ 0 \end{pmatrix}, \quad \begin{pmatrix} e_R \\ \\ -2 \end{pmatrix}. \tag{19.3}$$

The hypercharges are indicated in the bottom line (in the units of the coupling constant g'). One should not forget that u, d quarks carry also color indices. And the last useful remark is in order before we turn to particular anomalous triangles. The right-handed fermions propagating in the loop result in the anomalous contribution of the opposite sign compared to the left-handed fermions.

Now, after all the above preparations, we can verify the anomaly cancellation. The results for various triangle graphs of the type presented in Fig. 19.1 are as follows:

$\underline{Bgg}$

$$(u_L, d_L) \to 2 \times \frac{1}{3} \times \mathrm{Tr}(T^a T^b)_{\mathrm{color}},$$

$$u_R \to (-1)_{\mathrm{extra}} \frac{4}{3} \times \mathrm{Tr}(T^a T^b)_{\mathrm{color}},$$

$$d_R \to (-1)_{\mathrm{extra}} \left(-\frac{2}{3}\right) \times \mathrm{Tr}(T^a T^b)_{\mathrm{color}}$$

$$\text{total:} \to \left(\frac{2}{3} - \frac{4}{3} + \frac{2}{3}\right) \mathrm{Tr}(T^a T^b)_{\mathrm{color}} = 0\,. \tag{19.4}$$

$\underline{BBB}$

$$(u_L, d_L) \to 2 \times \left(\frac{1}{3}\right)^3 \times (3)_{\mathrm{color}},$$

$$u_R \to (-1)_{\mathrm{extra}} \left(\frac{4}{3}\right)^3 \times (3)_{\mathrm{color}}, \quad d_R \to (-1)_{\mathrm{extra}} \left(-\frac{2}{3}\right)^3 \times (3)_{\mathrm{color}}$$

$$(\nu_L, e_L) \to 2 \times (-1)^3,$$

$$e_R \to (-1)_{\mathrm{extra}} \times (-2)^3, \qquad \nu_R \to (-1)_{\mathrm{extra}} \times 0,$$

$$\text{total:} \to \frac{2}{9} - \frac{64}{9} + \frac{8}{9} - 2 + 8 + 0 = 0\,. \tag{19.5}$$

[3] For a good presentation of the Standard Model see e.g. Section 20.2 of Peskin and Schroeder or [1; 2].

$\underline{B W^3 W^3}$

$$(u_L,\, d_L) \to \frac{1}{3} \times (+1)^2 \times (3)_{\text{color}} + \frac{1}{3} \times (-1)^2 \times (3)_{\text{color}}\,,$$

$$(\nu_L,\, e_L) \to (-1) \times (+1)^2 + (-1) \times (-1)^2\,,$$

$$\text{total:} \to \; 1 + 1 - 1 - 1 = 0\,. \tag{19.6}$$

$\underline{B W^+ W^-}$

$$(u_L,\, d_L) \to \frac{1}{3} \times (+1)^2 \times (3)_{\text{color}} + \frac{1}{3} \times (+1)^2 \times (3)_{\text{color}}\,,$$

$$(\nu_L,\, e_L) \to (-1) \times (+1)^2 + (-1) \times (+1)^2\,,$$

$$\text{total:} \to \; 1 + 1 - 1 - 1 = 0\,. \tag{19.7}$$

$\underline{W^3 g g}$

$$(u_L,\, d_L) \to 1 \times \mathrm{Tr}(T^a T^b)_{\text{color}} - 1 \times \mathrm{Tr}(T^a T^b)_{\text{color}}\,,$$

$$\text{total:} \to \; 0\,. \tag{19.8}$$

$\underline{W^3 W^+ W^-}$

$$(u_L,\, d_L) \to 1 \times (+1)^2 \times (3)_{\text{color}} - 1 \times (+1)^2 \times (3)_{\text{color}}\,,$$

$$(\nu_L,\, e_L) \to 1 \times (+1)^2 + (-1) \times (+1)^2\,,$$

$$\text{total:} \to \; 0 + 0 = 0\,. \tag{19.9}$$

$\underline{W^3 W^3 W^3}$

$$(u_L,\, d_L) \to (+1)^3 \times (3)_{\text{color}} + \times (-1)^3 \times (3)_{\text{color}}\,,$$

$$(\nu_L,\, e_L) \to (+1)^3 + (-1)^3\,,$$

$$\text{total:} \to \; 0 + 0 = 0\,. \tag{19.10}$$

19.3 Peculiarity of the $SU(2)$ sector

Let us have a closer look at Eqs. (19.9) and (19.10) demonstrating cancellation of the triangle anomaly in the $SU(2)$ sector of the theory (i.e. the WWW anomaly). It is obvious that the cancellation pattern in this sector is not the same as in other triangles. Indeed, in the WWW case the anomaly is canceled not through assembling various multiplets, but rather inside each chiral multiplet separately. This is a special feature of $SU(2)$. In $SU(2)$ (and only in $SU(2)$) a chiral multiplet in the fundamental representation makes no contribution to the triangle anomaly. To understand the reason, let us consider the fact that the general anomalous triangle with three gauge-boson vertices in $SU(N)$ is proportional to

$$\mathrm{Tr}\left(T^a\left\{T^b T^c\right\}\right)_{SU(N)} \tag{19.11}$$

where T^a's are the generator matrices in the fundamental representation of $SU(N)$, and the braces denote anticommutator. Derivation of (19.11) can be carried out analogously to that in (19.9) and (19.10) if we assume that the generators to be used in (19.9) and (19.10) are not the Pauli matrices, but, rather, the generators of $SU(N)$.

The trace in (19.11) is proportional to the d^{abc} symbol, which vanishes for $SU(2)$ (but not for $SU(N)$ with $N > 2$.)

Nevertheless, the $SU(2)$ gauge theories in which the number of the fundamental fermion doublets is odd do not exist because of Witten's global anomaly.

SM was discovered by Glashow-Weinberg-Salam based on experimental data in the 1960s (see Appendix on page 205). Anomaly cancellation condition was not a factor in their analysis. And yet they – cancellations – occur. Extensive cancellations (19.4)–(19.10) obviously could not happen accidentally. They show that SM is very robust, and there is a deep and sophisticated underlying construction behind this model. One possible explanation is Grand Unification of all gauge groups.

19.4 Witten's global anomaly in SU(2) *

I won't be able to explain it in detail. A hint on its origin can be found in instanton calculus which we will briefly discuss later. An inspection of the instanton solution leads one to a perplexing question in SU(2) gauge theory. Indeed, let us assume that, we deal with one massless left-handed Weyl fermion transforming as a doublet with respect to SU(2). (Or it can be three, five and so on Weyl fermions in the fundamental representation of SU(2).)

In the SU(2) theory with a *single* massless left-handed Weyl fermion in the fundamental representation we will discover that the instanton-induced fermion vertex of the 't Hooft type must be *linear* in the fermion field. Indeed, in the instanton transition with one Dirac fermion $\Delta Q_5 = 2$, but the Weyl fermion $= 1/2$ of the Dirac fermion, and hence $\Delta Q_5 = 1$ (see Fig. 19.2).

It was obvious to many that something was wrong with this theory. The intuitive feeling of pathology was formalized by Witten who showed [3] that this theory is ill-defined because of a *global anomaly.* Such theory simply does not exist.

One of the possible proofs of the global anomaly is based on the fermion level restructuring in the instanton transition. The key elements are the following: (i) the vacuum-to-vacuum amplitude in the theory with one Weyl fermion is proportional to $\sqrt{\det(i\,D_\mu\gamma^\mu)}$; (ii) only one of the fermion levels changes its positions with regards to the sea level (I mean the Dirac sea, of course) when $\mathcal{K} = n$ goes in $\mathcal{K} = n+1$, as opposed to one *pair* in the case of the Dirac fermion (see pages 243 and 244). This forces the partition function to vanish making all correlation functions ill-defined.

19.5 Other anomalies *

Among other anomalies [4] I would like to mention the so-called *scale* anomaly which is intimately related to the generation of a dynamical scale parameter $\Lambda_{\rm QCD}$, see Eq. (14.18). We have already discussed the scale transformation in Sec. 16.2.1, page 175. Let us apply this transformation to the QCD action

$$S = \int d^4x \left\{ -\frac{1}{4}\left(G^{\mu\nu\,a}\,G^a_{\mu\nu}\right) + \sum_f \bar\psi_f\left(i\,\gamma^\mu D_\mu\right)\psi^f \right\}. \tag{19.12}$$

The quark masses is neglected here, see Eq. (2.1) for comparison. Classically, this expression is invariant under the scale transformations presented in Eq. (16.29) (plus $A \to \lambda^{-1}A$), see also Sec. 15.3. This invariance is in one-to-one correspondence with the statement of the

[4] Some of them are discussed in my course QFT III [4].

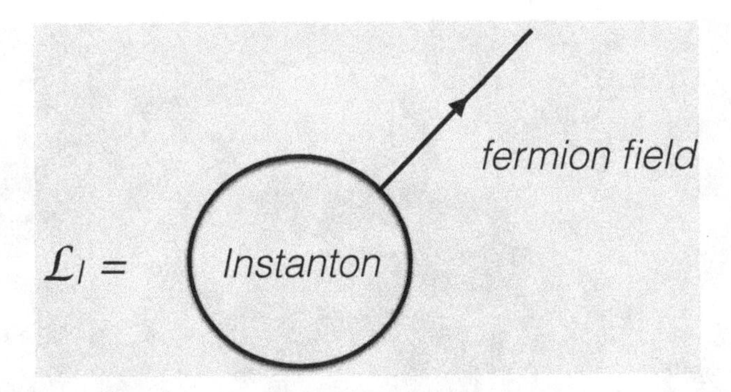

Fig. 19.2 Instanton-induced effective Lagrangian linear in the fermion field ($SU(2)$ gauge theory with one chiral fermion doublet).

tracelessness of the energy-momentum tensor which was considered in Sec. 15.3, Eq. (15.7). Classically the scale invariance of Eq. (19.12) implies that $\left(T^\mu_\mu\right)_{\rm class} = 0$. To see that in fact the scale symmetry in (19.12) is anomalous, one needs to regularize this expression. Derivation of the scale anomaly (a nonvanishing right-hand side in $T^\mu_\mu \propto G_{\mu\nu}G^{\mu\nu}$) becomes a very simple exercise, at least at one-loop level, if dimensional regularization is used. For lack of time, I refer the reader to [4], Sec. 36.

Appendix 19.1: Creators of the Standard Model

Sheldon Glashow, Abdus Salam, and Steven Weinberg were awarded the 1979 Nobel Prize in Physics for the creation of the Standard Model unifying the weak and electromagnetic interactions of quarks and leptons. The model also involves the Higgs mechanism for the mass generation of W and Z bosons. The existence of the electroweak interactions was experimentally established in two stages, the first being the discovery of neutral currents in neutrino scattering by the Gargamelle collaboration in 1973, and the second in 1983 by the UA1 and the UA2 collaborations that involved the discovery of the W and Z gauge bosons in proton–antiproton collisions at the converted Super Proton Synchrotron.

Sheldon Lee Glashow was born in 1932 in New York City, to Jewish immigrants from Russia, Bella (Rubin) and Lewis Gluchovsky, a plumber. He graduated from Bronx High School of Science in 1950. Glashow was in the same graduating class as Steven Weinberg. In 1961, Glashow extended electroweak unification models due to Schwinger by

Fig. 19.3 Sheldon Glashow.

including a short range neutral current, the Z_0. The resulting symmetry structure that Glashow proposed, $SU(2) \times U(1)$, forms the basis of SM. In collaboration with James Bjorken, Glashow was the first to predict a fourth quark, the charm quark, in 1964.

Glashow is the Metcalf Professor of Mathematics and Physics at Boston University and Higgins Professor of Physics, Emeritus, at Harvard University.

Steven Weinberg was born in 1933 in New York City, his parents were Jewish immigrants. He published famous works in such areas of the HEP theory as the high energy behavior of quantum field theory, spontaneous symmetry breaking, and pion scattering. After his 1967 seminal work on the unification of weak and electromagnetic interactions (which brought him the Nobel Prize), Steven Weinberg continued his work in quantum field theory, gravity, supersymmetry, superstrings and cosmology, as well as a theory called Technicolor. Currently Steven Weinberg holds the Josey Regental Chair in Science at the University of Texas at Austin. His research on elementary particles and cosmology has been honored with numerous prizes and awards, in addition to

Fig. 19.4 Steven Weinberg.

the 1979 Nobel Prize, namely the National Medal of Science in 1991, the Benjamin Franklin Medal of the American Philosophical Society in 2004, and so on. He has been elected to the US National Academy of Sciences and Britain's Royal Society, as well as to the American Philosophical Society and the American Academy of Arts and Sciences.

Mohammad Abdus Salam usually referred to as Abdus Salam (1926-1996) was a theoretical physicist of Pakistani descent. Salam was born to Chaudhry Muhammad Hussain and Hajira Hussain, into a Ahmadi Muslim Punjabi family. At age 14, Salam scored the highest marks ever recorded for the matriculation (entrance) examination at the Punjab University. In 1946, he was awarded a scholarship to St. John's College, Cambridge, UK, where he completed a BA degree with Double First-Class Honours in Mathematics and Physics in 1949. In 1950, he received the Smith's Prize from Cambridge University for the most outstanding pre-doctoral contribution to Physics. Later Abdus Salam renewed his scholarship from Cambridge where he defended his PhD thesis in 1951.

Fig. 19.5 Abdus Salam.

Salam made an important and significant contribution in quantum electrodynamics and quantum field theory, including its extension into particle and nuclear physics: he worked on the theory of neutrino, introduced the massive Higgs bosons to what later became the Standard Model, suggested $SU(2) \times U(1)$ gauge group for the theory of electroweak interactions, and contributed to the proof of the Goldstone's theorem. He was instrumental in the development of the superspace-based method in supersymmetric field theories.

In 1964, Salam founded the International Centre for Theoretical Physics (ICTP, currently named after Salam), Trieste, Italy and served as its director until 1993.

According to his biographers, Salam played a controversial role in Pakistani atomic bomb project.

References

[1] L. Okun, *Leptons and Quarks*, (World Scientific, Singapore, 2014).
[2] Stefan Pokorski, *Gauge Field Theories*, (Cambridge University Press; 2nd edition, 2000).
[3] E. Witten, *An SU(2) anomaly*, Phys. Lett. B **117**, 324 (1982).
[4] M. Shifman, *Advanced Topics in Quantum Field Theory*, (Cambridge University Press, 2012).

Chapter 20

Lecture 20. "External" Anomalies: Implications

The magic of the 't Hooft matching. Examples from QCD with massless quarks.

20.1 "External" anomalies

Let us now return to Chapter 18 with the aim of deriving crucial and highly valuable consequences. We will start by writing down the axial anomaly in QCD (weak interactions are switched off) in the triangle diagram for the axial current [1]

$$a_\mu \equiv j^5_\mu = \frac{1}{2}\left(\bar{u}\gamma_\mu\gamma^5 u - \bar{d}\gamma_\mu\gamma^5 d\right). \tag{20.1}$$

and two photons coupled to the electromagnetic current

$$j^{\rm em}_\mu = \frac{2}{3}\bar{u}\gamma_\mu u - \frac{1}{3}\bar{d}\gamma_\mu d\,. \tag{20.2}$$

The current (20.1) is external with regards to QCD. The photons will be considered as "external" as well. To distinguish them from the dynamical photon field in SM, we will temporarily denote the external photon field by $\mathcal{A}^\mu$. Then the interaction takes the form $e\mathcal{A}^\mu\, j^{\rm em}_\mu$. It is important that none of the currents above have "internal" anomalies in QCD. The vector current is anomaly free, of course. As for (20.1), for each of the terms $\bar{u}\gamma_\mu\gamma^5 u$ and $\bar{d}\gamma_\mu\gamma^5 d$ there are anomalous triangles with two gluons, but they cancel in the combination (20.1) because gluons are flavor blind.

[1] A more general formulation of the problem is as follows. Consider "strong" isospin, i.e. SU(2) symmetry generated by the currents $j^{5\,a}_\mu = \bar\Psi\frac{\tau^a}{2}\gamma_\mu\gamma^5\Psi$ and $j^a_\mu = \bar\Psi\frac{\tau^a}{2}\gamma_\mu\Psi$ where Ψ is a two-component column, with u and d placed up and down, respectively. By definition, there is no lepton contributions in the above currents. In the set of vector currents one can also include $j_\mu = \bar\Psi\frac{1}{2}\gamma_\mu\Psi$. Then one can consider all triangles that can be built from one axial-vector and two vector currents as defined in this footnote. The choice of the currents in (20.1) and (20.2) is a particular case. The number of colors N can be viewed as a free parameter.

All we have to do is to re-evaluate the diagram in Chapter 18 with the external gluons replaced by photons taking into account the difference in the vertex factors in this triangle graph. First, we will deal with the color factors, N ($N = 3$ in the real world). While in Chapter 18, see Eq. (18.10), with the gluon background field, we used $\mathrm{Tr_C}\left(T^a T^b\right) = \frac{1}{2}\delta^{ab}$, in the case of the photon background field we replace this by $\mathrm{Tr_C}\,1 = N$. Next, in the u loop we replace $g \to Q_u\, e$ and in the d loop $g \to Q_d\, e$. (Here $Q_u = 2/3$ and $Q_d = -1/3$.) As a result,

$$N_f\, g^2 \to \frac{1}{2}(2N)\left(Q_u^2 - Q_d^2\right) e^2\,, \tag{20.3}$$

where the factor $\frac{1}{2}$ is due to $\frac{1}{2}$ in the definition (20.1). Assembling all factors together we arrive at

$$\partial_\mu a^\mu = \frac{\alpha}{4\pi}\, N\left(Q_u^2 - Q_d^2\right) \mathcal{F}_{\mu\nu}\tilde{\mathcal{F}}^{\mu\nu} \tag{20.4}$$

where

$$\mathcal{F}_{\mu\nu} = \partial_\mu \mathcal{A}_\nu - \partial_\nu \mathcal{A}_\mu\,, \qquad \tilde{\mathcal{F}}^{\mu\nu} = \frac{1}{2}\varepsilon^{\mu\nu\alpha\beta}\mathcal{F}_{\alpha\beta}\,.$$

20.2 Longitudinal part of the current

Under certain circumstances one can reconstruct from (20.4) the longitudinal part of the current. Let us separate the longitudinal and transverse parts of a^μ,

$$a^\mu \equiv a^\mu_{||} + a^\mu_\perp\,, \qquad \partial_\mu a^\mu_\perp = 0\,. \tag{20.5}$$

It is clear that (20.4), viewed as an equation for the current, says nothing about $a^\mu_\perp$. However, it imposes a constraint on $a^\mu_{||}$, which allows one to unambiguously determine $a^\mu_{||}$ under appropriate kinematical conditions. Namely, assume that the photons in (20.4) are produced with the momenta $k^{(1)}$ and $k^{(2)}$ and are on mass shell, i.e.

$$\left[k^{(1)}\right]^2 = 0\,, \quad \left[k^{(2)}\right]^2 = 0\,. \tag{20.6}$$

The total momentum transferred from the current a^μ to the pair of photons is $q_\mu = k^{(1)}_\mu + k^{(2)}_\mu$ (see Fig. 20.1). Then

$$\mathcal{F}_{\mu\nu}\tilde{\mathcal{F}}^{\mu\nu} \longrightarrow -2 \times 2 \times \varepsilon^{\mu\nu\alpha\beta}\, k^{(1)}_\mu \epsilon^{(1)}_\nu\, k^{(2)}_\alpha \epsilon^{(2)}_\beta\,. \tag{20.7}$$

Here $\epsilon^{(1,2)}_\mu$ is the polarization vector of the first or second photon. The first factor of 2 in (20.7) comes from combinatorics: one can produce

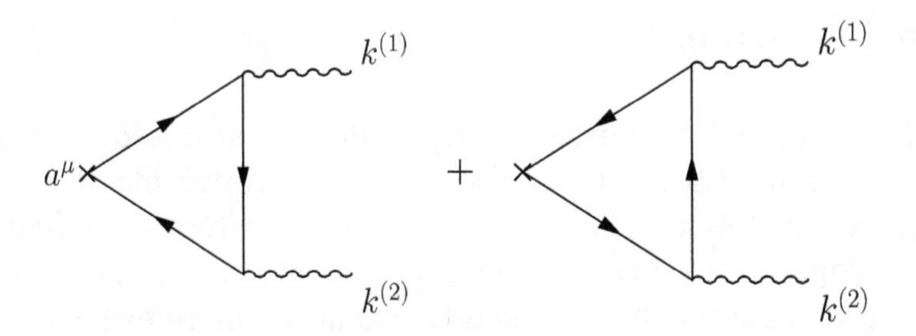

Fig. 20.1 Anomaly in a^μ. Note that if you go from the point a^μ along the fermion line flow in the diagram on the left, you first hit $k^{(1)}$ while in the diagram on the right you first hit $k^{(2)}$.

the first photon either from the first $\mathcal{F}_{\mu\nu}$ tensor or the second. Gauge invariance with regards to the external photons is built into the regularization.

The straightforward consequence from (20.4) *and* (20.6) is

$$\langle 0|\, a^\mu_{||}\, |2\gamma\rangle = i\,\frac{q^\mu}{q^2}\,\frac{\alpha}{\pi}\,N\left(Q_u^2 - Q_d^2\right)\varepsilon^{\mu\nu\alpha\beta}\,k^{(1)}_\mu \epsilon^{(1)}_\nu\, k^{(2)}_\alpha \epsilon^{(2)}_\beta\,. \qquad (20.8)$$

For on-mass-shell photons the two-photon matrix element of $a^\mu_{||}$ is determined unambiguously. This result is exact and is valid for any value of q^2, in particular, at $q^2 \to 0$. I would like to emphasize the emergence of the pole $1/q^2$, with far-reaching physical consequences.

That (20.8) is the solution to (20.4) is obvious. The fact that it is the *only* possible solution is less obvious and we will not discuss it here.

20.3 't Hooft matching and its physical implications

Now we will turn to physical consequences that can be obtained by using the so-called 't Hooft matching [1]. We will start from a general interpretation of the pole in (20.8) and similar anomalous relations for other currents, formulate the 't Hooft matching condition, argue (using large-N arguments) that the global $\mathrm{SU}(N_f)_A$ symmetry is spontaneously broken, and, finally, calculate the $\pi^0 \to 2\gamma$ decay width.

It is worth emphasizing that to derive the anomaly (20.4) we relied on an UV regularization. This is the UV-based result. On the other hand, the exact prediction of a pole in (20.8) is an IR statement. Thus, the anomalies reveal a connection between UV and IR physics which are exact in strongly coupled theories such as QCD. This can be viewed as a rare success!

Infrared matching

Poles do not appear in physical amplitudes for no reason. In fact, the only way an amplitude can acquire a pole is through massless particles in the spectrum of the theory coupled to the external currents under consideration. There are two possible scenarios: (i) the global axial symmetry is spontaneously broken (it would be more exact to say that it is realized nonlinearly); (ii) linear realization with massless spin-1/2 fermions.

In the first case, massless Goldstone bosons appear in the physical spectrum. They must be coupled to $j_\mu^{5,\,a}$ and external vector gauge bosons ("photons" in the case on hand). Equation (20.8) or similar equations for other currents present a constraint on the product of the Goldstone boson couplings which must and can be met.

The second scenario is more subtle and, apparently, is rather exotic. It is true that the triangle loop (Fig. 20.1) with massless spin-1/2 fermions yields $\frac{q^\mu}{q^2}$ in the longitudinal part $a_{||}^\mu$ of the axial current. However, not only is the kinematic factor $\frac{q^\mu}{q^2}$ exactly predicted by the anomaly, the coefficient in front of this factor is known *exactly* too. In QCD-like theories with confinement, for the chiral symmetry to remain unbroken, the massless spin-1/2 (composite, color-singlet) fermions that might be potential contributors to the triangle loop must exactly reproduce this coefficient, which, generally speaking, is a highly nontrivial requirement. The search for massless spin-1/2 fermions which could match the coefficient in front of $\frac{q^\mu}{q^2}$ in $a_{||}^\mu$ is one of the aspects of the celebrated 't Hooft matching procedure.

Needless to say, if free massless N-colored quarks existed in the spectrum of asymptotic states, they would automatically provide the required matching.[2] Alas ... quark confinement implies the absence of quarks in the physical spectrum. The only spin-1/2 fermions we deal with in QCD are composite baryons.

[2]In all theories strongly coupled in the infrared the only proper way of obtaining $a_{||}^\mu$ in the form (20.8) is the ultraviolet derivation through the external anomaly. However, if we pretend to forget all correct things about QCD and just blindly calculate the triangle loop of Fig. 20.1 with *noninteracting* massless quarks, we would get exactly the same formula. I hasten to add that this coincidence acquires a meaning only in the context of the 't Hooft matching. Feynman diagrams, in particular, that in Fig. 20.1, have no meaning whatsoever in the infrared in the language of QCD quark-gluon Lagrangian.

20.4 Spontaneous breaking of the axial symmetry

Let us see whether or not we can match (20.8) with the baryon contribution. We will put $N = 3$, as in our world, and consider first $N_f = 2$. Then the lowest-lying spin-1/2 baryons are proton and neutron (p and n), with the electric charges $Q_p = 1$ and $Q_n = 0$, respectively. Hence, only p contributes to the triangle loop in Fig. 20.1. If it were massless, it would generate a formula repeating (20.8) with the substitution

$$N\left(Q_u^2 - Q_d^2\right) \to Q_p^2\,. \tag{20.9}$$

The right- and left-hand sides in Eq. (20.9) are equal! Thus, in this particular case the 't Hooft matching does not rule out the linearly realized axial SU(2) symmetry with massless baryons p and n. This may be an accidental coincidence, though. Therefore, let us not make hasty conclusions and try to examine the stability of the above matching.

To this end we add the third quark, s, keeping intact the axial current to be analyzed, see (20.1). The electromagnetic current (20.2) acquires the additional term $-\frac{1}{3}\bar{s}\gamma_\mu s$. The UV-based anomaly prediction (20.8) remains intact.

In the theory with u, d and s quarks the lowest-lying spin-1/2 baryons form the baryon octet

$$B = (p,\, n,\, \Sigma^\pm,\, \Lambda, \Sigma^0,\, \Xi^-,\, \Xi^0). \tag{20.10}$$

If both the vector and axial SU(3) flavor symmetries are realized linearly, the baryon-baryon-photon coupling constants and the constants $\langle B|a^\mu|B\rangle$ at zero momentum transfer are unambiguously determined from the baryon quantum numbers (for instance, $\langle\Sigma^+|a^\mu|\Sigma^+\rangle = \bar{\Sigma}\gamma^\mu\gamma^5\Sigma$). Calculating the triangle diagram of Fig. 20.1 (more exactly, its longitudinal part) we find that the baryon octet does *not* contribute there due to cancellations: the proton contribution (the quark content uud) is canceled by that of Ξ^- (the quark content ssd) while the Σ^- contribution (the quark content dds) is canceled by Σ^+ (the quark content uus). Other baryons from (20.10) are neutral and decouple from the photon. The absence of matching seemingly tells us that the global $SU(3)_A$ symmetry must be spontaneously broken.

Although the above argument is suggestive, it is still inconclusive. It tacitly assumes that baryons with other quantum numbers, e.g. $J^P = \frac{1}{2}^-$, are irrelevant in the calculation of $a^\mu_{||}$, which need not be the case.

How can one argue that the combined contribution of all baryons cannot be equal to (20.8)?

To answer this question let us explore the N dependence in Eq. (20.8). The anomaly-based calculation naturally produces the factor N on the right-hand side. By the same token, saturating the triangle loop by baryons at large N, instead of the linear dependence on N, we would naturally arrive at the following statement [2]: each baryon loop is suppressed exponentially, as e^{-N}, since each baryon consists of N quarks and each quark loop brings in $1/N$ suppression. This argument leads us to the conclusion that the global $\mathrm{SU}(N_f)_A$ symmetry is spontaneously broken at least in the large-N limit. As a result, $N_f^2 - 1$ massless Goldstone bosons (pions) emerge in the spectrum. Note that this argument is inapplicable to the singlet axial current (see the remark at the end of Sec. 20.2). The singlet pseudoscalar meson need not be massless.

The assertion of the exponential suppression of the baryon loops has the status of a "physical proof" rather than a mathematical theorem. It is intuitively natural, indeed. However, in the absence of the full dynamical solution of Yang-Mills theories at strong coupling, one cannot completely rule out exotic scenarios in which the loop expansion in $1/N$ (implying e^{-N} for baryons) is invalid, see e.g. [3]. Alternatively one could say that an exponentially large number of baryons might contribute. Of course in QCD, experimental data confirm that this spontaneous breaking is the ultimate truth. However, doubts may remain concerning models with more contrived fermion sectors.

20.5 Predicting the $\pi^0 \to 2\gamma$ decay rate

If the global $\mathrm{SU}(N_f)_A$ symmetry is realized nonlinearly, through the Goldstone bosons (which for two flavors are called pions), the saturation of the anomaly-based formula (20.8) is trivial (Fig. 20.2). The pole in $a^\mu_{||}$ is due to the pion contribution. The constraint (20.8) provides us with a relation between the $a^\mu \to \pi^0$ amplitude and the $\pi^0 \to 2\gamma$ coupling constant. The result is known from the 1960s. For completeness I will recall its derivation.

The $\pi^0 \to \gamma\gamma$ amplitude can be parametrized as

$$A(\pi^0 \to 2\gamma) = F_{\pi 2\gamma}\, \mathcal{F}_{\mu\nu}\tilde{\mathcal{F}}^{\mu\nu} \to -4\, F_{\pi 2\gamma}\, k^{(1)}_\mu \epsilon^{(1)}_\nu\, k^{(2)}_\alpha \epsilon^{(2)}_\beta\, \varepsilon^{\mu\nu\alpha\beta}. \quad (20.11)$$

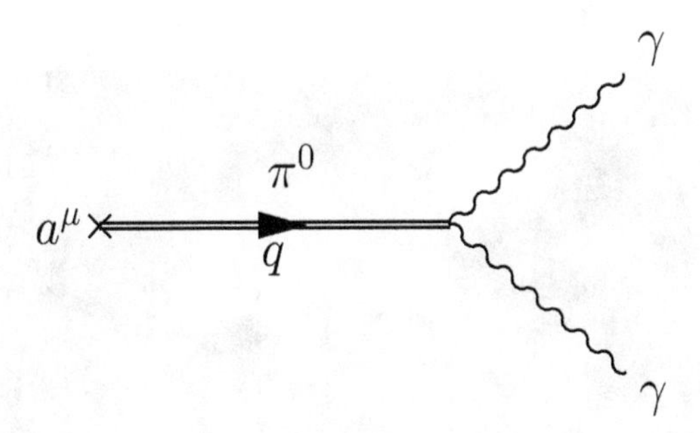

Fig. 20.2 The pion saturation of the anomaly.

Moreover, the amplitude $\langle 0|a^\mu|\pi^0\rangle$ is parametrized by the constant f_π playing the central role in pion physics,

$$\langle 0|a^\mu|\pi^0\rangle = \frac{1}{\sqrt{2}} i\, f_\pi\, q_\mu\,, \qquad f_\pi \approx 130\ \text{MeV}\,. \tag{20.12}$$

Then the pion contribution to the matrix element on the left-hand side of Eq. (20.8) is

$$\langle 0|\, a^\mu_{||}\, |2\gamma\rangle = i\,\frac{q^\mu}{q^2}\,\frac{f_\pi}{\sqrt{2}}\, 4\, F_{\pi 2\gamma}\, \varepsilon^{\mu\nu\alpha\beta}\, k^{(1)}_\mu \epsilon^{(1)}_\nu\, k^{(2)}_\alpha \epsilon^{(2)}_\beta\,. \tag{20.13}$$

Comparing with (20.8), we arrive at the following formula:

$$F_{\pi 2\gamma} = \frac{N}{2\sqrt{2}}\,\frac{1}{f_\pi}\,\frac{\alpha}{\pi}\left(Q_u^2 - Q_d^2\right) \to \frac{1}{2\sqrt{2}\, f_\pi}\,\frac{\alpha}{\pi}\,. \tag{20.14}$$

This is in good agreement with experiment.

Before the advent of QCD, people did not know about color; the factor $N = 3$ was omitted from the prediction (20.14). In fact, the analysis of the $\pi^0 \to \gamma\gamma$ decay was one of a very few quantitative proofs of existence of color in the early 1970s.

Appendix 20.1: Gerard 't Hooft

Gerardus (Gerard) 't Hooft (b. in 1946) is a Dutch theoretical physicist and professor at Utrecht University, the Netherlands. He shared the 1999 Nobel Prize in Physics with his thesis advisor Martinus J. G. Veltman "for elucidating the quantum structure of electroweak interactions." He authored some other breakthrough ideas, such as the so-called large-N 't Hooft limit and the 't Hooft-Polyakov monopole.

Fig. 20.3 Gerard 't Hooft.

References

[1] G. 't Hooft, *Naturalness, chiral symmetry, and spontaneous chiral symmetry breaking*, in *Recent Developments in Gauge Theories*, Eds. G. 't Hooft, C. Itzykson, A. Jaffe, H. Lehmann, P. K. Mitter, I. M. Singer and R. Stora, (Plenum Press, New York, 1980) [Reprinted in *Dynamical Symmetry Breaking*, Ed. E. Farhi *et al.*, (World Scientific, Singapore, 1982) p. 345 and in G. 't Hooft, *Under the Spell of the Gauge Principle*, (World Scientific, Singapore, 1994), p. 352].
[2] S. R. Coleman and E. Witten, *Chiral symmetry breakdown in large N chromodynamics*, Phys. Rev. Lett. **45**, 100 (1980).
[3] D. Amati and E. Rabinovici, Phys. Lett. B **101**, 407 (1981).

Chapter 21

Lecture 21. Divergence of Perturbation Theory at High Orders

Typically, perturbative expansions in field theories generate series in the powers of coupling constants with factorially divergent coefficients. We will discuss the physical meaning of this phenomenon. Dyson argument. Asymptotic series: what is it? Summing à la Borel.

In QFT-I and in this course (QFT-II) we learned how to calculate one-loop corrections in the coupling constant, mostly using gauge theories as examples, namely QED and Yang-Mills theories (QCD). In the 1950s, it was realized that expansions in the coupling constant α cannot represent conventional convergent series, with a finite radius of convergence. In the vast majority of cases the emerging series are *asymptotic*. This means that at first, as the expansion order grows, so does its accuracy. However, if we go too far in our perturbative expansion, typically to orders higher than $n = \text{const}/\alpha$, the accuracy of the series becomes worse rather than better. In other words, the expansion coefficients increase at high orders, and the perturbative series has vanishing convergence radius. Shortly, I will explain how this can happen. Right now, I explain why this is inevitable.

21.1 Dyson argument and factorial divergences?

Let us take QED for definiteness. Assume that we calculate some physical quantity A (it does not matter which particular quantity we consider) as an expansion in the powers of α,

$$A = (\text{a factor}) \times \left(1 + c_1\alpha^1 + c_2\alpha^2 + c_3\alpha^3 + ...\right). \tag{21.1}$$

We can choose any physical quantity which is real (i.e. has no imaginary part), for instance, the electron magnetic moment. Let us assume that the series (21.1) is convergent, with a finite convergent radius α_*. If so, it should remain convergent and define a real function of α for *negative* values of α at $|\alpha| < |\alpha_*|$. In the 1950s, Freeman Dyson noted that this cannot be correct. Indeed, were α negative, this would imply that the Coulomb interaction between electrons is attractive, while between electron and positron repulsive. In other words, this would "reverse" the electromagnetic interaction so that the same-sign charges would attract each other and the opposite-sign charges would repel.

If so, the vacuum (a "void") would be unstable against the production of a cluster of electrons at one point and another cluster of positrons far away from the former. This would become energetically expedient.

As usual, any instability with the subsequent decay gives rise to an imaginary part. Thus, we conclude that the initial assumption was wrong and the α series does not converge, but is in fact asymptotic. This consideration got the name of the "Dyson argument" (see [1]).

In more detail, one can explain the Dyson argument as follows.

Consider $2N$ charged moving particles, half positive, half negative, as depicted in Fig. 21.1, assuming that $R \gg r$ and $N \gg 1$. If average kinetic energy (including the mass term) is T, then each of the two separated clouds has kinetic energy TN. The potential energy stored in the cloud is

$$-\frac{N^2}{2}e^2\overline{r^{-1}}$$

where $\overline{r^{-1}}$ is the average inverse distance between the charges in each cloud. Note that because of the "reversal" mentioned above, the kinetic energy in the cloud is negative. The potential energy of the two-cloud interactions is positive but can be neglected provided $R \gg r$. Thus, the total energy is

$$2TN - N^2e^2\overline{r^{-1}}\,. \tag{21.2}$$

No matter how small $\alpha = e^2/4\pi$ is, the negative term dominates for sufficiently large N.

This instability is due to the fact that a spontaneous pair creation becomes energetically expedient if α is negative.

21.2 Asymptotic series and factorial divergences

The concept of asymptotic series: what is it?

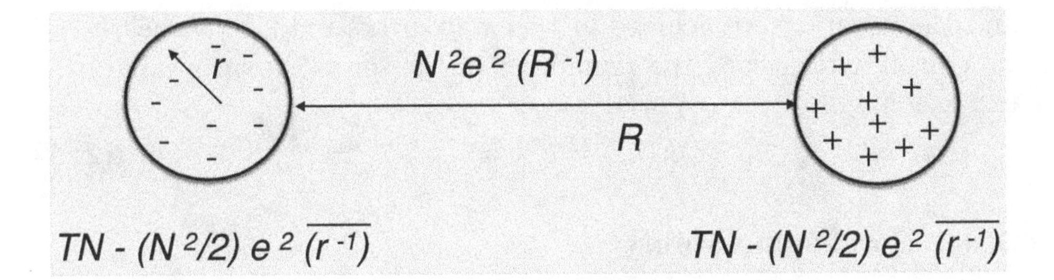

Fig. 21.1 Graphic representation of the Dyson instability argument.

The simplest example one can give is as follows. Consider the perturbative series of the form

$$Z = \sum_k C_k\, k!\, \alpha^k\,. \tag{21.3}$$

where all coefficients C_k are of order 1 and $k!$ stands for the factorial.[1] Here k is the number of loops. The coefficients in front of α^k grow factorially at large k; therefore, this is a divergent series. If the coefficients C_k are sign-alternating, the series is called Borel-summable, if these coefficients are of the same sign, the series is Borel-nonsummable. I will explain the meaning of this nomenclature shortly, see Sec. 21.3.

Why is it still useful?

Let us examine the k behavior of the subsequent terms in the series assuming that the expansion constant α is small. With k increasing $k!\alpha^k$ continues to fall until we reach $k = 1/\alpha$. Indeed, let us find the minimum in k using the Stirling formula for the factorial,

$$F(k) = k!\alpha^k \sim \left(\frac{k}{e}\right)^k \alpha^k = \exp\left(k \log \frac{k\alpha}{e}\right) \tag{21.4}$$

and

$$\frac{\partial F}{\partial k} = \left(\log \frac{k\alpha}{e} + 1\right) \exp\left(k \log \frac{k\alpha}{e}\right). \tag{21.5}$$

The right-hand side of (21.5) vanishes, i.e. the minimum of $F(k)$ is achieved at $k\alpha = 1$, implying that we can continue the above perturbative expansion up to

$$k_* = \frac{1}{\alpha}, \tag{21.6}$$

[1]The asymptotic divergence of the coefficients in QED is somewhat more contrived [2] than in (21.3) due to the fact that the QED loops are due to fermions.

and the accuracy we achieve in this way grows until k reaches k_*. If we cut off and discard the remainder of the series the accuracy of the expansion will be exponential,

$$F(k_*) = e^{-k_*} = e^{-1/\alpha}\,. \tag{21.7}$$

21.3 Borel summability

Instead of the *asymptotic series* (21.3), let us introduce the Borel transform defined as

$$B_Z(\alpha) \equiv \sum_k C_k \alpha^k\,. \tag{21.8}$$

In Eq. (21.8), the k-th term of expansion (21.3) is divided by $k!$, which implies, in turn, that the distance between the nearest singularity of $B_Z(\alpha)$ and the origin in the α plane is of order 1 (provided $C_k \sim 1$). Thus, the sum (21.8) is convergent, with the radius of convergence ~ 1.

Mathematicians would say that the function defined by (21.8) is obtained from (21.3) by the inverse Laplace transformation.

It is quite obvious that one can recover the original function Z performing the following integral transformation (the Laplace transformation):

$$Z(\alpha) = \int_0^\infty dt\, e^{-t} B_Z(\alpha\, t)\,, \tag{21.9}$$

Indeed, substituting (21.8) in (21.9) and integrating each term in the expansion separately we arrive at

$$\int_0^\infty dt\, e^{-t} \sum_k C_k\, t^k\, \alpha^k \tag{21.10}$$

and hence return to (21.3). The integral representation (21.9) is well-defined provided that $B_Z(\alpha)$ has no singularities on the real positive semi-axis in the complex α plane. This is the case if the asymptotic series (21.3) is sign-alternating, $C_k \sim (-1)^k$, (and then so is (21.8)). This is the case of Borel summability.

If however $B_Z(\alpha)$ has singularities on the real positive semi-axis (as is the case of all positive or all negative coefficients), then hitting the singularity of the integrand $B_Z(t\alpha)$, we face an ambiguity of order of $e^{-1/\alpha}$. One cannot resolve this ambiguity on the basis of purely mathematical arguments. More information is needed, which can be provided only by underlying physics. This is the Borel-nonsummable asymptotic series.

21.4 How can one determine the rate of factorial divergence

The answer to this question was first provided by A. Vainshtein [3] who realized how to convert the probability of the under-the-barrier penetration (the vacuum instability in the field theory language) for unphysical – negative – values of the expansion parameter into a prediction for the factorial divergence in (21.3). These two phenomena are in one-to-one correspondence with each other.

21.4.1 *The simplest quantum-mechanical example: Anharmonic oscillator*

Let us consider one-dimensional anharmonic oscillator,

$$\mathcal{H} = \frac{1}{2}p^2 + \frac{1}{2}\omega^2 x^2 + g^2 x^4 , \tag{21.11}$$

see Fig. 21.2. For definiteness we will focus on the ground-state energy E_0. There exists a well-defined procedure of constructing the expansion

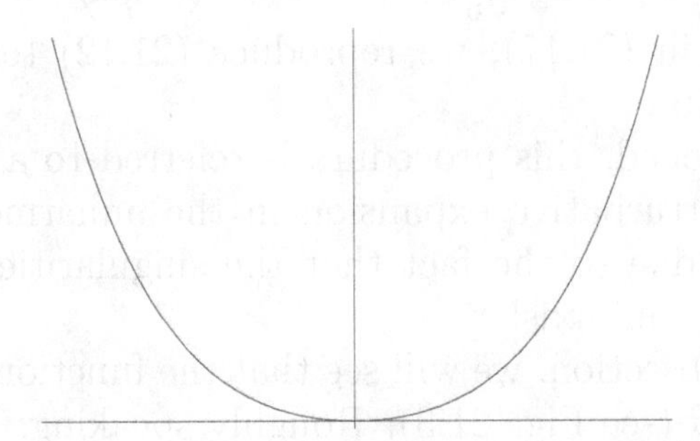

Fig. 21.2 $V(x)$ in the anharmonic oscillator problem (21.11).

for E_0 order by order in perturbation theory, to any finite order,

$$E_0 = \frac{\omega}{2}\left(1 + c_1 g^2 + c_2 g^4 + ...\right). \tag{21.12}$$

Nevertheless, Eq. (21.12) does not define the ground-state energy with arbitrary precision. Indeed, as was explained above, the coefficients c_k are factorially divergent at large k,

$$c_k \sim (-1)^k B^{-k} k! , \qquad k \gg 1 , \tag{21.13}$$

where

$$B = \frac{1}{3}\omega^3 \tag{21.14}$$

is proportional to the so-called bounce action, see Eq. (21.25).[2] As we will see shortly, it describes the tunneling probability from 0 to x^* in the potential in Fig. 21.4 Thus, the sum in (21.12) needs a regularization.

In the simplest case under consideration an appropriate (and exhaustive) regularization is provided by the Borel transformation $\mathcal{B}$,

$$\mathcal{B}E_0 \equiv \frac{\omega}{2}\left(1+\sum_{k=1}^{\infty}\frac{1}{k!}\,c_k\,g^{2k}\right) \equiv \frac{\omega}{2}\,f(g^2)\,. \tag{21.15}$$

The Borel transformation introduces $1/k!$ in each term of the series rendering it convergent. Moreover, if the convergent series

$$1+\sum_{k=1}^{\infty}\frac{1}{k!}\,c_k\,g^{2k} \equiv f(g^2) \tag{21.16}$$

defines a Borel function $f(a)$ with no singularities on the real positive semi-axis $a \geq 0$, then one can obtain the ground-state energy E_0 starting from the well-defined expression for $\mathcal{B}E_0$ and using the Laplace transformation,

$$E_0 = L\left(\mathcal{B}E_0\right) \equiv \frac{\omega}{2}\int_0^{\infty} da\, g^{-2}\exp\left(-\frac{a}{g^2}\right) f(a)\,. \tag{21.17}$$

Expanding $f(a)$ in (21.17), we reproduce (21.12) term by term, see (21.18) and (21.19).

As was mentioned, this procedure is referred to as the Borel summation. The perturbative expansion in the anharmonic oscillator is Borel-summable due to the fact that the singularities of $f(a)$ are on the *negative* real semi-axis!

In the next subsection, we will see that the function $f(a)$ indeed has a pole at $a = -B$ (see Fig. 21.3). Roughly speaking,

$$f(a) = \frac{B}{a+B}\,, \qquad B = \frac{\omega^3}{3}\,. \tag{21.18}$$

Then the integral (21.17) is well-defined. At the same time, expanding (21.18),

$$f(a) = \sum_{k=o}^{\infty}(-1)^k\left(\frac{a}{B}\right)^k, \tag{21.19}$$

and substituting this series in (21.17), we immediately arrive at (21.13). Equation (21.18) is confirmed in (21.24). Let me stress again that there is a one-to-one correspondence between two circumstances. The fact that the position of the singularity in the a plane is to the left of the origin entails sign alternation in (21.19). The opposite is true as well.

[2]Equation (21.13) is slightly simplified.

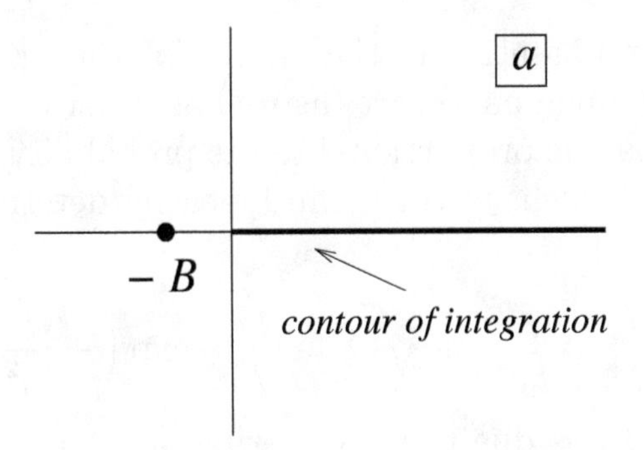

Fig. 21.3 The perturbative series in the anharmonic oscillator problem is Borel-summable. The g^2 series for E_0 is sign alternating; $f(a)$ has a singularity on the real negative semi-axis in the Borel parameter complex plane. a is the Borel parameter.

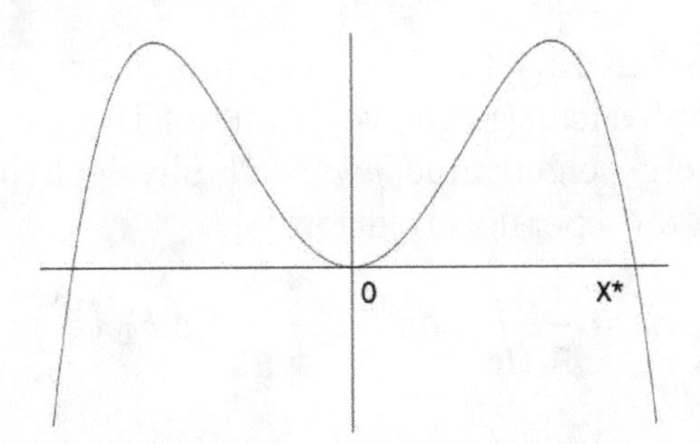

Fig. 21.4 The same potential with the replacement $g^2 \to -g^2$, to be denoted as $\tilde{V}(x)$.

21.4.2 *Tunneling*

Conceptually, our consideration is similar to the Dyson argument.

Changing the sign of g^2 from positive to negative, $g^2 \to -g^2$, one converts a stable potential $V(x)$ in (21.11) (see also Fig. 21.2) into an unstable potential $\tilde{V}(x)$,

$$\tilde{V}(x) = \frac{1}{2}\omega^2 x^2 - g^2 x^4 \tag{21.20}$$

presented in Fig 21.4. The latter allows for the leakage of the wave function from $x = 0$ to $x \geq x^*$ in Fig. 21.4. Here [3]

$$x^* = \frac{\omega}{\sqrt{2}\, g}\,. \tag{21.21}$$

[3] Of course, the wave function can leak from zero to $-x^*$ as well. This will only change a pre-exponential factor which we will not consider here.

In the leaking potential $\tilde{V}$ the energy of the ground state eigenvalue acquires an imaginary part (as well as other energy eigenvalues). This imaginary part is proportional to the probability of the under-the-barrier tunneling which, in turn, can be easily determined quasiclassically,

$$\Gamma_0 \sim \exp\left(-2\int_0^{x^*} dx\,\sqrt{2\tilde{V}(x)}\right) = \exp\left(-\frac{\omega^3}{3g^2}\cdot\right) \tag{21.22}$$

The first factor of 2 is due to the squaring of the amplitude. I ignored pre-exponential factors.

Now, we can write

$$\operatorname{Im} E_0 = \frac{\pi\,\omega}{2}\frac{B}{g^2}\exp\left(-\frac{B}{g^2}\right), \tag{21.23}$$

where B is defined in Eq. (21.18). The imaginary part of the energy eigenvalue for unphysical (negative) value of the coupling constant is related to the energy eigenvalue *per se* at physical (positive) coupling constant through a dispersion relation [3],

$$\begin{aligned} E_0 &= \frac{1}{\pi}\int_0^\infty d\tilde{g}^2\,\frac{1}{g^2+\tilde{g}^2}\operatorname{Im} E_0\left(\tilde{g}^2\right) \\ &= \frac{\omega}{2}\int dz\,\frac{1}{1+\frac{g^2}{B}z}\,e^{-z}\,. \end{aligned} \tag{21.24}$$

The last expression defines $f(a)$ as in Eq. (21.18).

In $g^2\phi^4$ field theory in four dimensions factorial divergence of the perturbative coefficients in high orders was calculated by L. Lipatov [4] who observed essentially the same behavior as in Eq. (21.13).

21.4.3 *Bounces*

The above consideration was quantum-mechanical – we calculated Γ_0 basing on WKB and the wave function, see (21.22). This method does not work in field theory. In order to calculate the tunneling probabilities in field theory, one can use either instantons or the so-called *bounces*, depending on the problem on hand. The bounce can be viewed as a version of instanton, though. The instantons are discussed in Chapter 22 and Appendix A. Let me say a few words about the bounces.

Inspecting the potential $\tilde{V}$ in Fig. 21.4, we immediately see that there is no classical trajectory connecting the points 0 and x^* in real time. However, performing the Euclidean rotation $t \to -it$ effectively changes the sign of the potential in the Euclidean action, as was explained in Chapter 3. After the Euclidean time rotation, the potential takes the form shown in Fig. 21.5.

What is the bounce trajectory? Assume we start at $x = 0$ and (infinitesimally) slightly push the system to the right. It starts rolling down, passes the minimum and continues the motion until it reaches the point $x = x^*$. At this point its velocity vanishes, then reverses the sign, the system starts rolling down, then up, until it returns to the point $x = 0$ stopping at rest. At a certain moment of time it bounces off the "wall" which is reflected in its name, the bounce. Thus, in Euclidean, the bounce trajectory is classically accessible.

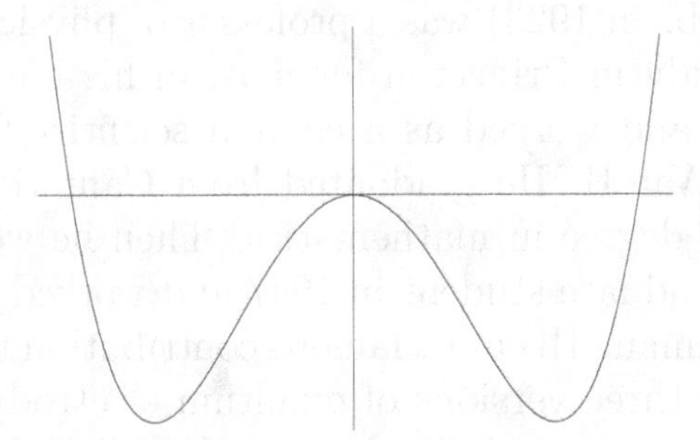

Fig. 21.5 An effective potential in Euclidean time. This potential presents a sign reflection of that in Fig. 21.4, i.e. $-\tilde{V}(x)$. It vanishes at $x = 0$ and at $x = \pm x_*$ where $x_* = \frac{\omega}{\sqrt{2}\,g}$.

The Euclidean action on the bounce trajectory $A_{\rm bounce}$ is readily calculable through the corresponding classical equation of motion. It is easy to see that

$$A_{\rm bounce} = \frac{B}{g^2}\,, \tag{21.25}$$

where B is defined in Eq. (21.18).

Summarizing, the perturbative expansion for the anharmonic oscillator is factorially divergent; however, the Borel summability allows one to find the closed, well-defined and exact expressions for the energy eigenvalues. The physical meaning of the factorial divergence, as well as the sign alternation, are fully understood.

*****

A more challenging but more interesting case of Borel non-summable perturbative series would have something like

$$\frac{1}{1 - \frac{g^2}{B} z}$$

in the second line of (21.24). Discussion of the Borel non-summable situation goes beyond the scope of this lecture course.

Appendix 21.1: On Freeman Dyson

Freeman Dyson (b. in 1923) was a professor of physics at the Institute for Advanced Study in Princeton for most of his life. He was born in England in 1923 and worked as a civilian scientist for the Royal Air Force in World War II. He graduated from Cambridge University in 1945 with a B.A. degree in mathematics. Then he went on to Cornell University as a graduate student in 1947 and worked with Hans Bethe and Richard Feynman. His most famous contribution to physics was the unification of the three versions of quantum electrodynamics invented by Feynman, Schwinger and Tomonaga. Cornell University made him a professor without bothering about his lack of PhD.

Fig. 21.6 Freeman Dyson: left, at retirement; right, in 2016 (age 92).

References

[1] F. J. Dyson, *Divergence of perturbation theory in quantum electrodynamics*, Phys. Rev. **85**, 631 (1952) [Reprinted in *Large-Order Behavior of Perturbation Theory*, ed. J.C. Le Guillou and J. Zinn-Justin (North-Holland, Amsterdam, 1990).

[2] E. B. Bogomolny and V. A. Fateev, Phys. Lett. B **76**, 210 (1978).

[3] A. I. Vainshtein, *Decaying Systems and Divergence of the Series of Perturbation Theory*, Novosibirsk preprint, 1964. Its English translation was first published in *Continuous Advances in QCD 2002*, ArkadyFest, ed. K. Olive, M. Shifman, and M. Voloshin (World Scientific, Singapore, 2002), p. 619; see also https://www.academia.edu/17865243/Arkady_Vainshtein_40_Year_Journey_in_Theoretical_Physics, page 1.

[4] L. N. Lipatov, Sov. Phys. – JETP **45**, 216 (1977).

Chapter 22

Lecture 22. Degenerate (Pre)Vacua and Instantons in QM *

> *"Prevacua" in quantum mechanics and field theory. Instantons and their role. Rarified instantons at weak coupling.*

22.1 A brief review of a well-known QM problem

Let us start from an example which most of you probably remember from the course of quantum mechanics (QM): double well potential with two degenerate wells separated by a large barrier. I will review its "standard" solution, good in all respects except it cannot be generalized to QFT. I will then outline the instanton solution generalizable to QFT, following which we will discuss instantons in Yang-Mills theories.

A generic double well potential is depicted in Fig. 22.1. Two symmetric wells (I and II in Fig. 22.1) are separated by a barrier which is large at weak coupling. In classical physics this barrier is non-penetrable for particles with energy below the barrier height. If this were the case in QM, there would exist two degenerate vacua corresponding to oscillations in domains I and II, and the Z_2 symmetry of the model (obvious from Fig. 22.1) would remain unbroken. This is impossible in quantum mechanics. Tunneling from I to II or vice versa will "mix" the wave function even if the initial condition places the particle, say in I. Correspondingly, two energy eigenlevels will appear: one symmetric and another antisymmetric under

$$x \to -x\,, \tag{22.1}$$

and their energies will no longer be degenerate, they will split.

The standard calculation of the energy splitting for the ground state is as follows.

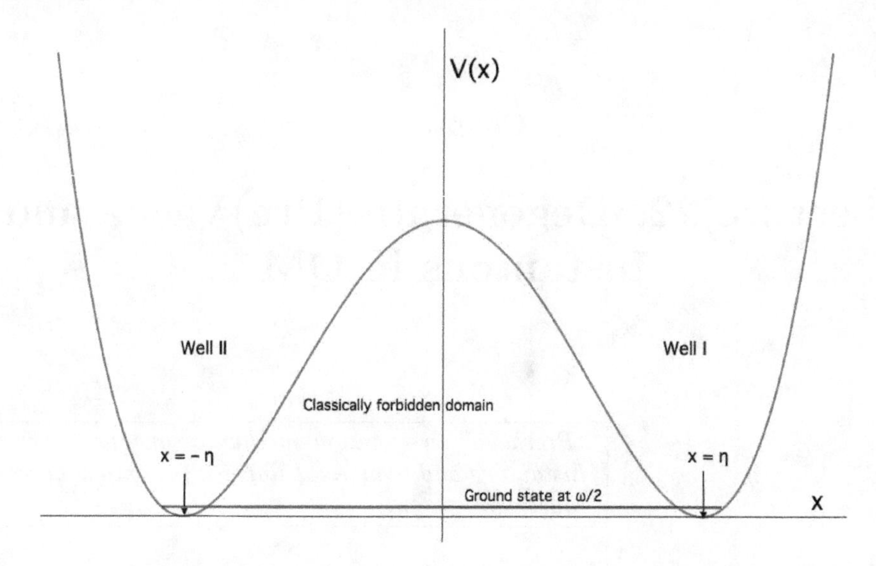

Fig. 22.1 Double well potential $U(x)$ the horizontal red line at $\omega/2$ indicates the ground state energy of a harmonic oscillator approximating the bottom of the graph either in the domain I or the domain II.

Let E_0 be the energy eigenvalue corresponding to the quantum wave function localized in, say, well I. Denote the corresponding localized wave function as $\psi_0(x)$ assuming that the normalizing integral over the well I is unity,

$$\int_I (\psi_0(x))^2 \, dx = 1 \,. \tag{22.2}$$

Then, the wave function localized at II will be $\psi_0(-x)$, with the same energy eigenvalue E_0. Since at weak coupling the barrier penetration is exponentially suppressed, E_0 will be split $E_0 \to E_{1,2}$. The symmetric and antisymmetric wave functions are

$$\psi_1(x) = \frac{1}{\sqrt{2}}\left(\psi_0(x) + \psi_0(-x)\right), \qquad \psi_2(x) = \frac{1}{\sqrt{2}}\left(\psi_0(x) - \psi_0(-x)\right). \tag{22.3}$$

Quasiclassically, $\psi_0(x)$ dies off away from I while $\psi_0(-x)$ dies off away from II. The Schrödinger equations can be written as

$$\psi_0'' + 2(E_0 - U)\psi_0 = 0 \,, \qquad \psi_1'' + 2(E_1 - U)\psi_1 = 0 \,, \tag{22.4}$$

where the mass is taken to be $m = 1$. Next, we multiply the first equation by ψ_1 and the second by ψ_0 and subtract one from another. The result of this subtraction can be integrated over dx in the interval from 0 to ∞. We will need the following equations:

$$\int_0^\infty \psi_0\,\psi_1\,dx \approx \frac{1}{\sqrt{2}}\int_0^\infty \psi_0\,\psi_0\,dx = \frac{1}{\sqrt{2}} \tag{22.5}$$

and

$$\int_0^\infty dx\left(\psi_1\psi_0'' - \psi_0\psi_1''\right) = \left(\psi_1(x)\psi_0'(x) - \psi_0(x)\psi_1'(x)\right)_0^\infty, \tag{22.6}$$

where in (22.6) we do integration by parts. At the upper limit $\psi_1(x)\psi_0'(x) - \psi_0(x)\psi_1'(x)$ vanishes. Therefore,

$$\int_0^\infty dx\left(\psi_1\psi_0'' - \psi_0\psi_1''\right) = -\frac{1}{\sqrt{2}}\left(\psi_0(-x)\psi_0'(x) - \psi_0(x)\psi_0'(-x)\right)\Big|_0$$

$$= -\sqrt{2}\,\psi_0(0)\,\psi_0'(0)\,. \tag{22.7}$$

Combining the above expressions we arrive at

$$E_1 - E_0 = -\psi_0(0)\psi'(0)\,. \tag{22.8}$$

The approximate equality in (22.5) is valid up to exponentially small terms.

Using the quasiclassical approximation for $\psi_0(x)$ we find

$$\psi_0(0) = \sqrt{\frac{\omega}{2\pi v_0}}\,\exp\left(-\int_0^a |p|dx\right), \tag{22.9}$$

where $v_0 = \sqrt{2(U_0 - E_0)}$ and $\omega = 2\pi/T$ is the frequency of the classical oscillation near the bottom of I (find it for the potential $U = \lambda(x^2 - \eta^2)^2$).

Finally, in summary, the general formula for the energy splitting is

$$E_2 - E_1 = \frac{\omega}{\pi}\,\exp\left(-\int_{-a}^a |p|dx\right), \tag{22.10}$$

where p is the momentum in the classically forbidden domain. Strictly speaking, the exponent in Eq. (22.10) depends on energy (i.e. E_0 or E_1). However, if we normalize the potential at its minima as

$$U_{\rm min} = 0\,, \tag{22.11}$$

one can replace E_0 in the exponent by zero. This will not change the exponential suppressing factor but will show up in the pre-exponent, of which we are not so much interested now. This is because $\omega \ll U(0)$, the height of the barrier.

The last question we may ask ourselves is what happens if instead of two degenerate minima we would have n equidistant minima, say, at $x_1, x_2, \dots x_n$ (with $n \gg 1$)? The limiting case which we will have in mind is a periodic function with $n = \infty$.

Then we will have n prevacua: $\psi_0(x-x_1)$, $\psi_0(x-x_2), \dots, \psi_0(x-x_n)$. The ground state and its family – instead of two levels in (22.3) – will have n levels with exponentially small energy splittings,

$$\Psi_n(x) \approx \sum_{m=1}^{n} \left[\psi_0(x - x_m) \exp\left(\frac{2\pi i\, m\, k}{n}\right) \right]. \tag{22.12}$$

When we pass from one minimum to another we multiply ψ by

$$\exp\left(\frac{2\pi i\, k}{n}\right), \tag{22.13}$$

where $k = 0, 1, 2, \dots n-1$ is an integer. In the limit $n \to \infty$, the quantity $2\pi k/n$ becomes a continuous parameter θ varying from 0 to 2π.

22.2 Instantons in QM

In order to get an idea on how one can treat such problems in QFT where the notions of "wave function" and the Schrödinger equation are rather useless, let us consider path integral method applied to the double well potential QM problem. We will choose a specific parametrization of the potential in Fig. 22.1, namely,

$$U(x) = \lambda(x^2 - \eta^2)^2, \tag{22.14}$$

where

$$8\lambda\eta^2 = \omega^2 \tag{22.15}$$

and we assume that $\lambda \ll \omega^3$ so that the theory is weakly coupled and the barrier is indeed large. Near the positive minimum of the potential

$$U(x) \approx \frac{\omega^2}{2}(x - \eta)^2. \tag{22.16}$$

Since we are interested in the impact of tunneling we must choose the boundary conditions appropriately, namely,

$$x(t \to -\infty) = -\eta, \qquad x(t \to \infty) = +\eta, \tag{22.17}$$

i.e. the trajectory starts in the left classical minimum of the potential in the distant past and ends in the right classical minimum of the potential in the distant future. It is quite obvious that in the Minkowski time there is no such classical trajectory. Hence, lesson number one: to study tunneling in the path integral formalism, we must pass to the Euclidean time,

$$t \to -i\tau\,. \tag{22.18}$$

Then

$$iS[x(t)]_{\rm Mink} \to -S[x(\tau)]_{\rm Eucl}\,, \qquad S_{\rm Eucl} = \int \left[\frac{1}{2}\left(\frac{dx}{d\tau}\right)^2 + U(x)\right]. \tag{22.19}$$

The classical equation of motion takes the form

$$\frac{d^2x(\tau)}{d\tau^2} = U'(x)\,. \tag{22.20}$$

We can immediately reinterpret (22.20) as a classical motion in the potential $-U$, see Fig. 22.2. Now it is obvious that the minimal action trajectory satisfying the boundary conditions (22.17) exists, and, moreover, its analytical form is known,

$$x_*(\tau) = \eta \tanh\frac{\omega(\tau-\tau_0)}{2}\,, \tag{22.21}$$

where the subscript c stands for classical, and τ_0 is a free parameter called the instanton center. This solution is depicted in Fig. 22.3.

[Prove that Eq. (22.15) with the boundary conditions (22.12) imply that instead of (22.15) you can look for the solution of the first order equation $dx(\tau)/d\tau = -\sqrt{2\lambda}(x^2-\eta^2)$.]

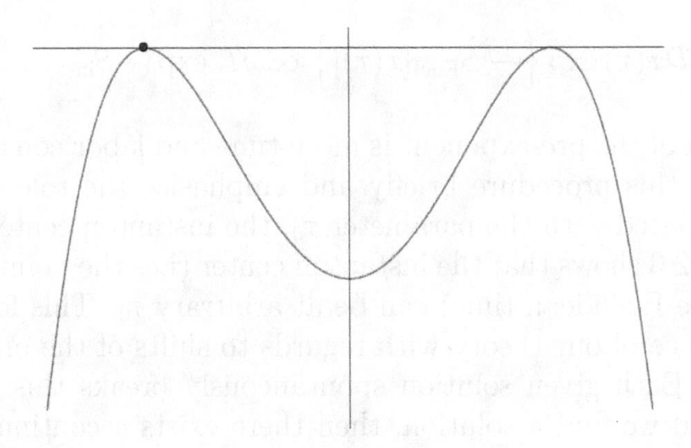

Fig. 22.2 Effective potential for the solution of the classical equation of motion in Euclidean time, see Eq. (22.20).

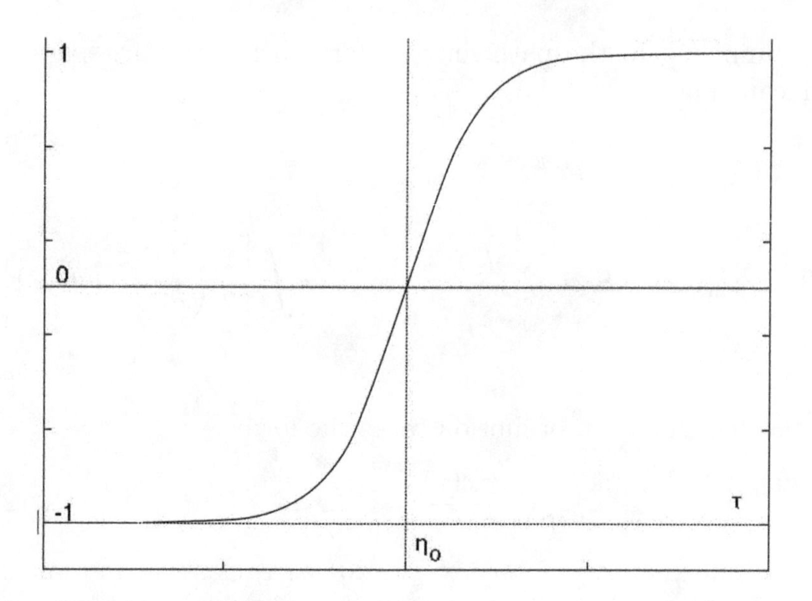

Fig. 22.3 The trajectory $x_*(\tau)/\eta$ in the Euclidean time connecting the vacua II and I with the minimal action.

The action corresponding to the solution (22.16) is

$$S_{\text{inst}} = \int d\tau \frac{dx_*(\tau)}{d\tau} = \frac{\omega^3}{12\lambda}\,. \tag{22.22}$$

Thus, the amplitude of the transition from $x_i = -\eta$ in the distant past (i.e. at $\tau = -T/2$ where $T \to \infty$ at the end of our calculation) to $x_f = \eta$ in the distant future (i.e. at $\tau = T/2$) is given by the functional integral[1]

$$\int_{x_i}^{x_f} \mathcal{D}x(\tau) \exp\Big\{ - S_{\text{Eucl}}[x(\tau)]\Big\} \propto \omega T \, \exp\left(-S_{\text{inst}}\right)\,. \tag{22.23}$$

Calculation of the pre-exponent is more time and labor consuming. We will review this procedure briefly and emphasize the role of the zero mode associated with the parameter τ_0, the instanton center.

Figure 22.3 shows that the instanton center (i.e. the point of passage $x = 0$ in the Euclidean time) can be at arbitrary τ_0. This follows from the invariance of our theory with regards to shifts of the origin on the time axis. Each given solution spontaneously breaks this invariance. Therefore, if we find a solution, then there exists a continuous family of solutions with the center point shifted arbitrarily. It is intuitively

[1]For the explanation of the emergence of the pre-exponential factor ωT, see below.

obvious that in calculating the instanton contribution to the transition with the boundary conditions (22.17) we must integrate over all τ_0. Thus the exponent (22.23) must be multiplied by $\int d\tau = T$. This factor is not the end of story, however.

In general, as we know, the pre-exponent is obtained if we include small fluctuations near the trajectory x_* in our calculation of the path integral. In other words, we write

$$x(\tau) = X_*(\tau) + \sum_n c_n x_n(\tau) \qquad (22.24)$$

where $x_n(\tau)$ is a complete infinite set of orthonormal functions which vanish at the boundaries

$$x_n(\pm T/2) = 0\,, \qquad \int_{-T/2}^{T/2} d\tau\, x_m(\tau) x_n(\tau) = \delta_{mn}\,. \qquad (22.25)$$

Substituting (22.24) in the action we obtain

$$S[x(\tau) + \delta x(\tau)] = S_0 + \int d\tau \delta x \left[-\frac{1}{2}\frac{d^2}{d\tau^2}\,\delta x + \frac{1}{2} U''(x_*)\,\delta x \right] \qquad (22.26)$$

where the absence of terms linear in δx is explained by the fact that x_* is the solution of the classical equation of motion. From (22.26) it is clear that the most convenient choice of the basis of x_n's is formed by the eigenfunctions of the operator

$$-\frac{d^2}{d\tau^2} + U''(x_*)\,. \qquad (22.27)$$

Then the path integral obviously takes the form

$$\int_{x_i}^{x_f} \mathcal{D}x(\tau) \exp\Big\{ - S_{\text{Eucl}}[x(\tau)] \Big\}$$

$$= \exp\left(-S_{\text{inst}}\right) \times \left[\det{}' \left(-\frac{d^2}{d\tau^2} + U''(x_*) \right) \right]^{-1/2}. \qquad (22.28)$$

The prime attached to the det sign means that if there are zero modes, they should be removed, otherwise the zero mode contribution to (22.28) would be $1/\sqrt{0}$, which is clearly ill-defined.

In fact, there is a zero mode, associated with the time translation invariance. The action for any η_0 in the solution (22.21) is the same; this action is η_0 independent. Therefore the derivative

$$\frac{d}{d\tau_0}\, x_*(\tau - \tau_0) \qquad (22.29)$$

is the zero mode. The corresponding coefficient c_0 in (22.24) does not enter the action. This explains the removal of the zero mode from the determinant in (22.28). Integration over dc_0 in the path integral is divergent, and should be replaced by $\omega \int d\tau_0 = \omega T$ in the pre-exponent. At $T \to \infty$ the factor ωT in the pre-exponent is also divergent.

22.3 Topological charge

In the given model the topological charge Q is defined as

$$Q = \frac{1}{2\eta} \int_{-\infty}^{\infty} \dot{x}\, dt\,. \tag{22.30}$$

It is clear that for *any* trajectory which starts and ends in one and the same pre-vacuum we have $Q = 0$, while if a trajectory starts in one pre-vacuum and ends in the other (i.e. interpolates between the two distinct pre-vacua)

$$Q = \pm 1\,. \tag{22.31}$$

This is due to the fact that the integrand in (22.30) is a full derivative. The unity in (22.31) is guaranteed by an appropriate normalization factor in (22.30).

22.4 Instanton gas

If, instead of a single instanton we consider a rarified gas of arbitrary number n of instantons and sum over n, this will exponentiate the one-instanton expression (22.23), and we will arrive at the partition function

$$Z = e^{-E_0 T} + \ldots = \exp\left[-\frac{\omega T}{2} \pm \omega T \exp\left(-S_{\text{inst}}\right)\right] + \ldots \tag{22.32}$$

where on both sides of (22.32) only the ground state contribution is retained. The plus or minus sign on the right-hand side of (22.32) is correlated with the fact that we can have either odd or even number of instantons in our ensemble. The term $-\frac{\omega T}{2}$ in the square brackets is due to the perturbative contribution (the ground state energy in the harmonic oscillator is $\omega/2$).

Equation (22.32) implies that the ground state energy in the double-well potential of Fig. 22.1 (which would be degenerate in the absence of tunneling) is split into two levels: one symmetric, and another antisymmetric under the reflection of x with respect to the origin,

$$E_0 = \frac{\omega T}{2} \mp \omega T \exp\left(-S_{\rm inst}\right) + \dots . \tag{22.33}$$

The symmetric level has lower energy than the antisymmetric but the splitting is exponentially small at weak coupling (i.e. if $S_{\rm inst} \gg 1$, which is the necessary and sufficient condition of our quasiclassical approximation). The result (22.33) coincides with that which can be obtained by the standard quantum-mechanical method of the wave function matching [1].

***

Students often ask why the boundary conditions in the problem of the double-well potential are chosen as in Eq. (22.17). For each given problem the choice of the appropriate boundary conditions is a matter of convenience – one chooses them on physical grounds in such a way as to arrive at the final result in the fastest way. In a different problem, other choices are more relevant. For instance, in calculating the tails of the wave functions far away from classically forbidden domains the most expedient boundary conditions were found in [2].

References

[1] L. D. Landau and E. M. Lifshitz, *Quantum Mechanics: Non-Relativistic Theory*, Third Edition, (Elsevier, 1977), page 183, Problem 3.

[2] M. A. Escobar-Ruiz, E. Shuryak and A. V. Turbiner, *Quantum and thermal fluctuations in quantum mechanics and field theories from a new version of semiclassical theory,* Phys. Rev. D **93**, no. 10, 105039 (2016).

Appendix A

Lecture 23. Nontrivial Topology in the Space of Fields *

This and the previous lecture on page 229 are complementary. Usually semester ends before I am done with the previous material.

Instantons in Yang-Mills

Yang-Mills instantons are localized objects in four-dimensional Euclidean space-time, with finite action. Originally Polyakov suggested the name "pseudoparticles," which did not take root, however, and is used rather rarely. The term instantons was suggested by 't Hooft. The physical role of instantons is as follows: in the quasiclassical approximation they describe the least-action trajectory (in the Euclidean time) which connects two distinct energy-degenerate states in the space of fields. The initial point of the instanton trajectory at $t = -\infty$ is one such state, while the final point at $t = \infty$ is another such state. Naturally, instantons are present only in those theories in which energy-degenerate states in the space of fields exist. They minimize the (Euclidean) action, under the given boundary conditions. Therefore, instantons present classical solutions of the Euclidean equations of motion. In fact, as we will see shortly, they satisfy the so-called duality equations. In non-Abelian gauge theories they were discovered by Belavin, Polyakov, Schwarz, and Tyupkin and are usually referred to as BPST instantons.

In this course we will consider only pure Yang-Mills theory with the gauge group SU(2). [In QCD the gauge group is SU(3).] Inclusion of fermions and the passage from SU(2) to SU(3) are beyond this course.

Nontrivial topology in the space of Yang-Mills fields

The Yang-Mills Lagrangian has the form

$$\mathcal{L} = -\frac{1}{4g^2} G^a_{\mu\nu} G^a_{\mu\nu} \tag{A.1}$$

where $G_{\mu\nu}$ is the gluon field strength tensor,

$$G^a_{\mu\nu} = \partial_\mu A^a_\nu - \partial_\nu A^a_\mu + f^{abc} A^b_\mu A^c_\nu \,, \tag{A.2}$$

g is the gauge coupling constant, f^{abc}'s stand for the structure constants. For SU(2)

$$f^{abc} = \varepsilon^{abc} \qquad a, b, c = 1, 2, 3\,.$$

For the time being there is no need for gauge fixing.

The first question to be asked is from where to where does the system of the Yang-Mills fields tunnel.

At first glance it is not obvious at all that the Lagrangian (A.1) has a discrete set of degenerate classical minima, or, as we called them in QM problem, *pre-vacua.* But it does!

The space of fields in field theories is infinitely-dimensional. Most of these field-theoretical degrees of freedom are oscillator-like and, thus, having just a single ground state, present no interest for our current purposes. However, we will see that in Yang-Mills theories there exists one composite degree of freedom, a direction in the infinitely-dimensional space of fields along which the Yang-Mills system can tunnel. If we forget for a while about all other degrees of freedom, and focus on this chosen degree of freedom, we will see degenerate states connected by "under-the-barrier" trajectories.

The closest analogy one can keep in mind while analyzing Yang-Mills theories in the context of tunneling is quantum mechanics of a particle living on a vertically oriented circle subject to a constant gravitational force (Fig. A.1). Classically, the particle with the lowest possible energy (the ground state of the system) just stays at rest at the bottom of the circle. Quantum-mechanically, the zero-point oscillations come into play. Within the perturbative treatment, we deal exclusively with small oscillations near the equilibrium point at the bottom of the circle. For such small oscillations, the existence of the upper part of the circle plays no role. It could have been eliminated altogether with no impact on the zero-point oscillations.

From the courses of quantum mechanics we know, however, that the genuine ground-state wave function is different. The particle oscillating

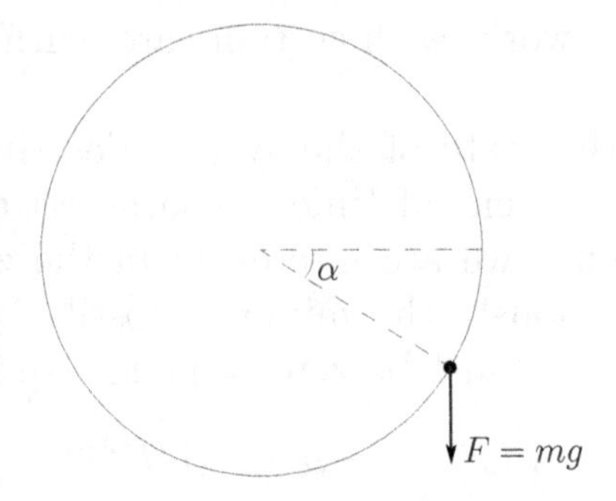

Fig. A.1 Quantum mechanics of a particle on a one-dimensional topologically nontrivial manifold, the circle.

near the origin "feels" that it can wind around the circle to which it belongs, through tunneling under the potential barrier it experiences at the top of the circle (the barrier is shown in Fig. A.2).

To single out a relevant degree of freedom in the infinitely-dimensional space of the gluon fields, it is necessary to proceed to the Hamiltonian formulation of the Yang-Mills theory which implies, of course, that the time component of the four-potential A_μ has to be gauged away, $A_0 = 0$. Then,

$$\mathcal{H} = \frac{1}{2}\int d^3x \, \{E_i^a E_i^a + B_i^a B_i^a\} \tag{A.3}$$

where $\mathcal{H}$ is the Hamiltonian, and $E_i^a = \dot{A}_i^a$ are to be treated as canonical momenta.

Two subtle points are to be mentioned in connection with this Hamiltonian. First, the equation $\text{div}\vec{E}^a = \rho^a$, inherent to the original Yang-Mills theory, does not stem from this Hamiltonian *per se*. This equation must be imposed by hand, as a constraint on the states from the Hilbert space. Second, the gauge freedom is not fully eliminated. Gauge transformations which depend on $\vec{x}$ but not t are still allowed. This freedom is reflected in the fact that, instead of two transverse degrees of freedom $\vec{A}_\perp^a$, the Hamiltonian above has three (three components of $\vec{A}^a$). Imposing, say, the Coulomb gauge condition,

$$\partial_i A_i^a = 0 \tag{A.4}$$

we could get rid of the "superfluous" degree of freedom, a procedure quite standard in perturbation theory (in the Coulomb gauge). Alas! If we want to keep and reveal the topologically nontrivial structure of the space of the Yang-Mills fields, the Coulomb gauge condition *cannot* be imposed. We have to work, with certain care, with the "undergauged" Hamiltonian.

Quasiclassically the state of the system described by Hamiltonian (A.3) at any given moment of time is characterized by the field configuration $A_i^a(\vec{x})$. Since we are interested in the zero-energy states – classically, this is obviously the minimal possible energy – the corresponding gauge field A_i must be zero or gauge equivalent to zero,

$$A_i(\vec{x})\Big|_{\text{vac}} = iU(\vec{x})\,\partial_i U^\dagger(\vec{x}) \tag{A.5}$$

where U is a matrix belonging to SU(2) and depending on the spatial components $\vec{x}$.

Moreover, we are interested only in those zero-energy states which may be connected with each other by tunneling transitions, i.e. the corresponding classical action must be finite. The latter requirement results in the following boundary condition: [1]

$$U(\vec{x}) \to 1, \qquad |\vec{x}| \to \infty\,, \tag{A.6}$$

or any other constant matrix U_0 independent of the direction in the three-dimensional space along which $\vec{x}$ tends to infinity. This boundary condition *compactifies* our three-dimensional space which thus becomes topologically equivalent to three-dimensional sphere S_3.

On the other hand, the group space of SU(2) is also a three-dimensional sphere. Indeed, any matrix belonging to SU(2) can be parametrized as

$$M = A + i\vec{B}\vec{\tau}, \qquad M \in \text{SU}(2)\,. \tag{A.7}$$

Here A and $\vec{B}$ are four real parameters; $\vec{\tau}$ are the Pauli matrices. Both conditions, $M^+M = 1$ and $\det M = 1$, are met, provided

$$A^2 + \sum_{i=1}^{3} B_i = 1\,. \tag{A.8}$$

Since $U(\vec{x})$ is a matrix from SU(2), and the space of all coordinates $\vec{x}$ is topologically equivalent to a three-dimensional sphere (after compactification $U(\vec{x}) \to 1$ at $|\vec{x}| \to \infty$), the function $U(\vec{x})$ realizes a mapping of the sphere in the coordinate space onto a sphere in the group space. Intuitively, it is quite obvious that all continuous mappings $S_3 \to S_3$ are classified according to the number of coverings, i.e.

[1] If (A.6) is not satisfied, then $G_{0i} \sim \dot{A}_i$ at large fixed t will scale as $1/|\vec{x}|$ and the integral $\int d^3x\, G_{0i}^2$ will be divergent implying infinite action.

the number is the number of times we sweep the group-space sphere S_3 when the coordinate $\vec{x}$ sweeps the sphere in the coordinate space once. The number of coverings can be zero (a topologically trivial mapping), one, two, and so on. The number of coverings can be negative too, since the mappings $S_3 \to S_3$ are orientable. Mathematically this is expressed by the formula

$$\pi_3(S_3) = \mathbb{Z}\,. \tag{A.9}$$

In other words, the matrices $U(\vec{x})$ can be sorted out in distinct classes labeled by an integer number, $U_n(\vec{x})$, $n = 0, \pm 1, \pm 2, \ldots$, which is referred to as the *winding number.* All matrices belonging to a given class $U_n(\vec{x})$ are reducible to each other by a continuous $\vec{x}$-dependent gauge transformation. At the same time, no continuous gauge transformation can transform $U_n(\vec{x})$ into $U_{n'}(\vec{x})$ if $n \neq n'$. The unit matrix represents the class $U_0(\vec{x})$. For $n = 1$ one can take, for instance,

$$U_1(\vec{x}) = \exp\left[-i\pi \frac{\vec{x}\vec{\tau}}{(\vec{x}^2 + \rho^2)^{1/2}}\right], \tag{A.10}$$

where ρ is an arbitrary parameter. An example of the matrix from U_n is $[U_1(\vec{x})]^n$.

Any field configuration $A_i(\vec{x})|_{\rm vac} = iU_n(\vec{x})\partial_i U_n^\dagger(\vec{x})$, being pure gauge, corresponds to the lowest possible energy – the zero energy. As a matter of fact, the set of points $\{U_n\}$ in the space of fields consists simply of the gauge images of one and the same physical point (analogous to the bottom of the circle in Fig. A.1). The fact that the matrices U_n from different classes are not continuously transformable to each other indicates the existence of a "hole" in the space of fields, with noncontractible loops winding around this "hole."

We are finally ready to identify the degree of freedom which "lives" on a circle. Define

$$K^\mu = 2\varepsilon^{\mu\nu\alpha\beta}\left(A_\nu^a \partial_\alpha A_\beta^a + \frac{g}{3} f^{abc} A_\nu^a A_\alpha^b A_\beta^c\right), \quad \varepsilon^{0123} = 1\,. \tag{A.11}$$

The vector K^μ is called the *Chern-Simons current.* We will encounter it more than once in what follows. Now, define the charge $\mathcal{K}$ corresponding to the Chern-Simons current,

$$\mathcal{K} = \frac{g^2}{32\pi^2}\int K_0(x)\, d^3x\,. \tag{A.12}$$

It is not difficult to show that for any pure gauge field $A_i^a(\vec{x})$ the Chern-Simons charge $\mathcal{K}$ measures the winding number. For any field of type (A.5) we have

$$\mathcal{K} = n\,. \tag{A.13}$$

Summarizing, moving in the "direction of $\mathcal{K}$" in the space of the Yang-Mills fields, we observe that this particular direction has the topology of the circle. The points $\mathcal{K}$ and $\mathcal{K} \pm 1$, and $\mathcal{K} \pm 2$, and so on are physically one and the same point. The integer values of $\mathcal{K}$ correspond to the bottom of the circle in Fig. A.1.

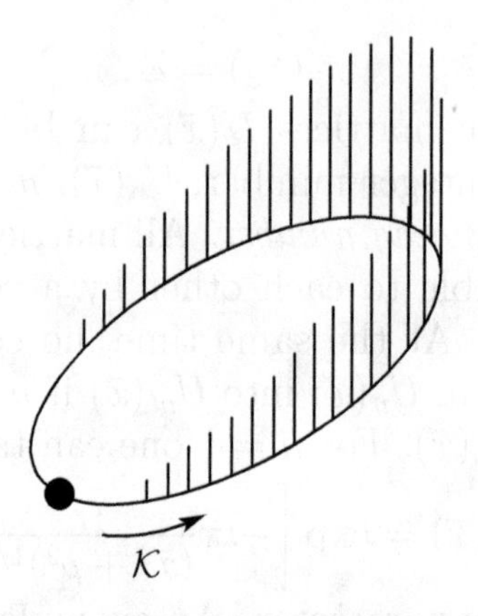

Fig. A.2 Nontrivial topology in the space of gauge fields in the $\mathcal{K}$ direction. The length of the circle is 1. The vertical lines indicate the strength of a potential acting on the effective degree of freedom living on the circle.

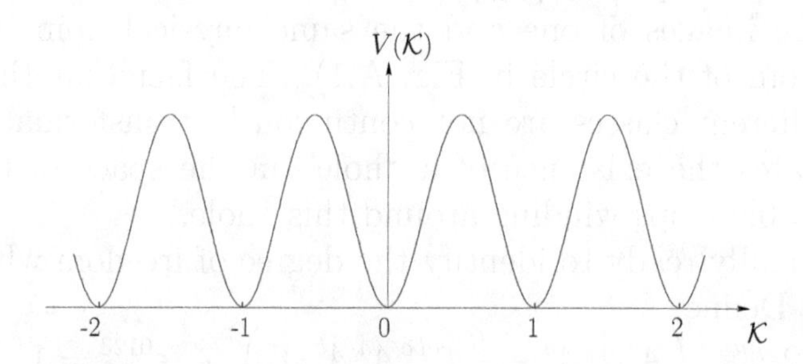

Fig. A.3 If we unwind the circle of Fig. A.2 onto a line we get a periodic potential.

It is convenient to visualize the dynamics of the Yang-Mills system in the "direction of $\mathcal{K}$" as shown in Fig. A.2. The vertical lines indicate the potential energy – the higher the line the larger the energy. It is well-known that the only consistent way of treating quantum-mechanical systems living on a circle (angle-type degrees of freedom) is to cut the circle and map it many times onto a straight line. In other words, we pretend that the variable $\mathcal{K}$ lives on the line (Fig. A.3). Any integer value of $\mathcal{K}$ in Fig. A.3 corresponds to a pure gauge configuration with zero energy. On the other hand, if $\mathcal{K}$ is not an integer, the field strength tensor is nonvanishing and the energy of the field configuration is positive. Viewed as a function on the line, the potential energy $V(\mathcal{K})$ is, of course, periodic – with the unit period.

To take into account the fact that the original problem is formulated on the circle, we impose the (quasi)periodic Bloch boundary condition on the wave functions Ψ,

$$\Psi(\mathcal{K}+1) = e^{i\theta}\,\Psi(\mathcal{K})\,. \tag{A.14}$$

The phase θ (note: $0 \le \theta \le 2\pi$) appearing in the Bloch quasiperiodic boundary condition is a hidden parameter, the *vacuum angle.* The boundary condition (A.14) must be one and the same for the wave functions of all states. We will return to the issue of the vacuum angle later on. The classical minima of the potential in Fig. A.3 can be called *pre-vacua.* The correct wave function of the quantum-mechanical vacuum state of the Bloch form is a linear combination of these pre-vacua.

We would like to emphasize here a subtle point which in many presentations remains fogged. It might seem that the systems depicted in Figs. A.2 and A.3 (a particle on a circle and that in the periodic potential) are physically identical. This is not quite the case. In the periodic potentials, say in crystals, one can always introduce impurities that would slightly violate periodicity. For a system on the circle this cannot be done. The correct analog system for Yang-Mills theories, where the gauge invariance is a sacred principle, is that of Fig. A.2.

Assume that at $\tau = -\infty$ and at $\tau = +\infty$ our system is at the classical minima (zero-energy state) depicted in Fig. A.3, but the minimum in the past is different from that in the future. Assume that at $\tau = -\infty$ the winding number $\mathcal{K} = n$ while at $\tau = +\infty$ the winding number $\mathcal{K} = n \pm 1$. In Fig. A.2, this means that our system tunnels from the point marked by the closed circle to the very same point under the hump of the potential energy.

Consider now a field configuration $A_\mu(\tau, \vec{x})$ continuously interpolating (with the minimal action) between these two states in the Euclidean time, the least-action tunneling trajectory. This is the Belavin-Polyakov-Swartz-Tyupkin (BPST) instanton.

The analysis outlined above (the one based on the Hamiltonian formulation) is convenient for establishing the existence of a nontrivial topology, nonequivalent (pre)vacuum states and, hence, the existence of nontrivial interpolating field configurations corresponding to tunneling. In practice, however, the Hamiltonian gauge $A_0 = 0$ is rarely used in constructing the instanton solutions. This gauge is inconvenient for this purpose.

Below we will describe a standard procedure based on a specific *ansatz* for $A_\mu(x)$ in which all four Lorentz components of A_μ are nonvanishing. This *ansatz* entangles the color and Lorentz indices; the field configurations emerging in this way are, following Polyakov, generically referred to as "hedgehogs."

Theta vacuum and θ term

The existence of a noncontractible loop in the space of fields A_μ leads to drastic consequences for the vacuum structure in non-Abelian gauge theories. Let us take a closer look at the potential of Fig. A.3. The argument presented below is formulated in the quasiclassical language. One should keep in mind, however, that the general conclusion is valid even though the quasiclassical approximation is inappropriate in quantum chromodynamics where the coupling constant becomes large at large distances.

The lowest-energy state of the system depicted in Fig. A.3, classically, is in one of the minima of the potential. Quantum-mechanically the zero point oscillations arise. The wave function [2] corresponding to oscillations near the n-th zero-energy state, Ψ_n, is localized near the corresponding minimum. The genuine wave function is delocalized, however, and takes the form

$$\Psi_\theta = \sum_{n=0,\pm1,\pm2,\ldots} e^{in\theta}\,\Psi_n\,, \tag{A.15}$$

where θ is a parameter,

$$0 \le \theta \le 2\pi\,, \tag{A.16}$$

analogous to the quasimomentum in the physics of crystals. If the n-th term in the sum is an n-th "pre-vacuum," the total sum represents the θ vacuum. The vacuum angle θ is a global fundamental constant characterizing the boundary condition on the wave function. It does not make sense to say that in one part of the space θ takes some value, while in another part θ takes a different value, or that θ depends on time.

[2] In application to Yang-Mills theories and QCD, we should rather use the term wave functional; nevertheless, we will continue referring to the wave function.

Once this parameter is set, we stay in the world with the given θ vacuum forever. The wave functions with different values of θ are orthogonal. More exactly, for any operator $\mathcal{O}$ from the Hilbert space of the physical states

$$\langle \Psi_\theta | \mathcal{O} | \Psi_{\theta'} \rangle = 0 \quad \text{if} \quad \theta \neq \theta' \,. \tag{A.17}$$

This property is referred to as the *superselection rule.*

The energy of Ψ_θ can (and does) depend on θ, generally, and so do other physically measurable quantities. From the definition of the vacuum angle, it is clear that the θ dependence of all physical observables, including the vacuum energy, must be periodic, with the period 2π.

Although the physical meaning of the parameter θ is absolutely transparent within the Hamiltonian formulation, usually, when we speak of instantons in field theory, we keep in mind the Lagrangian formulation based on the path-integral formalism. In the Lagrangian formalism, the vacuum angle is introduced as a θ term in the Lagrangian (cf. $G_{\mu\nu}\widetilde{G}_{\mu\nu} = \partial_\mu K_\mu$)

$$\mathcal{L} = -\frac{1}{4} G^a_{\mu\nu} G^{a,\,\mu\nu} + \mathcal{L}_\theta, \qquad \mathcal{L}_\theta = \theta \frac{g^2}{32\pi^2} G^{a,\,\mu\nu} \widetilde{G}^a_{\mu\nu} \,, \tag{A.18}$$

where

$$\widetilde{G}^a_{\mu\nu} = \frac{1}{2} \varepsilon_{\mu\nu\alpha\beta} G^{a,\,\alpha\beta} \,, \qquad \varepsilon^{0123} = 1 \,. \tag{A.19}$$

Note that if $\theta \neq 0$ or π, the θ term violates P and T invariance.

Before the discovery of the instantons, it was believed that QCD naturally conserves P and CP. Indeed, the only gauge invariant Lorentz scalar operator one could construct from the A_μ fields of dimension 4 violating P and T is $G\widetilde{G}$. This operator, however, presents a full derivative, $G\widetilde{G} = \partial_\mu K_\mu$ where K_μ is the Chern-Simons current (A.11). It was believed that such a full derivative has no impact on the action.

In the instanton field, however, the integral over $G\widetilde{G}$ does *not* vanish. The reasons for that will be explained below. What is important for us now is the fact that adding the θ term to the QCD Lagrangian, we do break P and CP in the strong interactions if $\theta \neq 0$ or π. Since it is known experimentally, that P and CP symmetries are conserved in the strong interactions, to a very high degree of accuracy, this means that in nature the vacuum angle is fine-tuned, and is very close to zero.[3] Estimates show that $\theta \lesssim 10^{-9}$.

[3]The second solution, with $\theta = \pi$, does not go through.

Thus, with the advent of instantons the naturalness of QCD is gone. Can this fine-tuning be naturally explained? There exist several suggestions as to how one could solve the problem of P and CP conservation in QCD in a natural way. One of the most popular is the axion conjecture. This topic, however, definitely lies outside the scope of this course.

In the Minkowski space the θ term is real. It becomes purely imaginary in passing to the Euclidean space. Certainly, this does not mean any loss of unitarity.

The transition from the Minkowski to Euclidean formulation is discussed in detail in Sec. 2.3, see page 17. If you plan to read the next lecture it is absolutely necessary to return to Sec. 2.3 to refresh your knowledge of passing to Euclidean space.

Appendix B

Problems

All problems marked by asterisk are not required for your home assignment. These are extra problems which can bring you 10 extra points each.

Problem 1 (L1)

Write down the generators of the $SU(3)$ group in the fundamental representation. Use the standard form of the Gell-Mann matrices. Commute them and find the set of structure constants for $SU(3)$. The generators T^a are defined through the commutation relation $[T^a T^b] = i f^{abc} T^c$.

Problem 2 (L1)

Determine the mass dimension of the constant h in (1.38) at $D = 4$. Can you do it for $D = 2$ and 3?

Solution

The general formula is

$$\dim h = \frac{4 - D}{2}.$$

Problem 3* (L1)

For $SU(N)$ group with any N find the generators in the *adjoint* representation in terms of the structure constants.

Solution

Dimension of the generator matrices in the adjoint representation is $(N^2-1)\times(N^2-1)$. Let us show that these matrices can be chosen as follows:

$$(T^a)_{AB} = if^{AaB}\,, \tag{B.1}$$

where the small Latin letter marks the number of the generator while the capital Latin letters are the matrix indices.

To prove Eq. (B.1), we must take a step back and use the Jacobi identity valid for any three square matrices P, Q, and R:

$$[P[QR]] + [Q[RP]] + [R[PQ]] = 0\,. \tag{B.2}$$

Next, replace in (B.2)

$$P \to T^A\,, \quad Q \to T^a\,, \quad R \to T^b \tag{B.3}$$

and use the general definition of the structure constants,

$$[T^aT^b] = if^{abD}T^D\,. \tag{B.4}$$

In this way, one arrives at the following identity for the structure constants:

$$f^{abB}\,f^{ABD} + f^{bAB}\,f^{aBD} + f^{AaB}\,f^{bBD} = 0\,. \tag{B.5}$$

Sometimes, it is referred to as the Jacobi identity for the structure constants.

Now, let us return to our conjecture (B.1). Substituting (B.1) into the defining relation (B.4) we get

$$f\times f - f\times f = f\times f \tag{B.6}$$

with the appropriate indices for each f. If we explicitly write down these indices and compare the result with the identity (B.5) we will see that (B.4) is indeed satisfied.

Problem 4 (L2)

Assume we have Yang-Mills theory with the gauge sector

$$G = SU(2)\,, \tag{B.7}$$

and the Higgs sector consisting of one real scalar field in the *adjoint* representation of $SU(2)$.

Assuming that the adjoint field develops a vacuum expectation value

$$\langle\phi^a\phi^a\rangle = v^2 \tag{B.8}$$

(large in the scale of quantum corrections) determine the masses of the gauge bosons in terms of the above expectation value v.

Note: The model above is the so-called Georgi-Glashow model. It was designed as a replacement of the Glashow-Weinberg-Salam model in the early days of the latter, when it was still unclear whether or not GWS was supported by data. Now it is widely used in other applications.

*Problem 4a**. More complicated question: Do the same for the gauge theory with $G = SU(3)$.

Hint: First find the simplest form to which one can reduce a generic adjoint field by using $SU(3)$ rotations. How many (independent) real quantities parametrize this generic form?

Solution

In the $SU(2)$ case any arbitrary x independent vacuum field ϕ can be reduced to the diagonal form using the gauge transformations, $\phi \equiv \phi^a T^a \to U \phi U^\dagger$ where the generator matrix is taken in the fundamental representation. This is due to the fact that all generators are Hermitean matrices. Then

$$\phi = \frac{v}{2}\begin{pmatrix} 1 & 0 \\ 0 & -1 \end{pmatrix}. \tag{B.9}$$

where the overall normalization follows from (B.8). Equation (B.9) presents the most general form and corresponds to $\phi^3 = v\,, \phi^1 = \phi^2 = 0$ where v is real.

As we already know, the mass term for the gauge bosons is determined from the kinetic term of the ϕ fields assuming that $\phi^a = (\phi^a)_{\text{vac}} = \{0, 0, v\}$. Then

$$\frac{1}{2}(D^\mu \phi^a)(D_\mu \phi^a) \to \frac{1}{2} g^2 v^2 \varepsilon^{ab'3} A_\mu^{b'}\, \varepsilon^{ab3} A^{\mu\, b} \to \frac{1}{2} g^2 v^2 \left[\left(A_\mu^1\right)^2 + \left(A_\mu^2\right)^2 \right] \tag{B.10}$$

implying that the mass of A_μ^3 vanishes (the analogue of the photon). Thus the $U(1)$ subgroup of $SU(2)$ remains un-Higgsed. Instead of $A_\mu^{1,2}$, we introduce the combinations which are analogous to W bosons,

$$W^\pm = \frac{1}{\sqrt{2}} \left(A_\mu^1 \pm i A_\mu^2\right). \tag{B.11}$$

From (B.10) we see that the masses of the "W bosons" are gv. The pattern of the gauge symmetry breaking is $SU(2) \to U(1)$. It is instructive to count components. The field ϕ^a has three real components. With the above pattern of the symmetry breaking, two scalar fields become the longitudinal components of $W^{\pm}$, while the third one (oscillations of ϕ^3 is the physical Higgs particle.

Now let us pass to $SU(3)$. The adjoint field ϕ in this case is a 3×3 matrix,

$$\phi = \frac{1}{2}\phi^a \lambda^a , \tag{B.12}$$

where λ^a are eight Gell-Mann matrices.

Using the gauge freedom $\phi \equiv \phi^a T^a \to U\phi U^\dagger$ we can always reduce $(\phi)_{\rm vac}$ to a diagonal and traceless form,

$$(\phi)_{\rm vac} \sim \begin{pmatrix} a\, 0 & 0 \\ 0\, b & 0 \\ 0\, 0 & -a-b \end{pmatrix}, \tag{B.13}$$

where a and b are two independent real parameters. This means that

$$\phi^3 \neq 0, \quad \phi^8 \neq 0, \quad \phi^{1,2,4,5,6,7} = 0 . \tag{B.14}$$

If we repeat the calculation of the mass terms, as above, we will find that the mass terms are *not* acquired by A^3_μ and A^8_μ, while all other gauge bosons generally acquire unequal masses. The corresponding pattern of the symmetry breaking is $SU(3) \to U(1) \times U(1)$. You may want to calculate the masses of six massive gauge bosons in the simplest case $a = -b$.

Problem 5 (L2)

Assume we have $SU(N)$ Yang-Mills theories with different types of massless fermions. Namely, two theories are to be considered, with two distinct types of fermions.

In the first theory, we deal with one Dirac fermion in the fundamental representation of $SU(N)$. In the second theory, the fermion sector contains one Majorana fermion in the adjoint representation of $SU(N)$.

WARNING: Do NOT perform full calculation of the diagram below!

Find the ratio of the diagrams above in the two given cases (for arbitrary N). What happens with this ratio if $N \to \infty$?

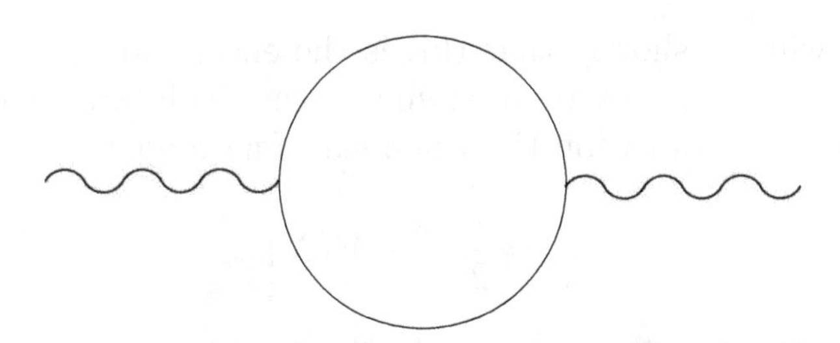

Fig. B.1 The fermion loop contribution to the gluon Green function. The wavy lines represent gauge fields (gluons), the solid line represents fermions. Two theories are to be considered, with two distinct type of fermions. The momentum of the incoming and outgoing gauge field line is q, where $q^2 < 0$.

Solution

Two factors distinguish the former case from the latter. The diagram above is proportional to $\text{Tr}T^aT^b$ which is equal to $\frac{1}{2}\delta^{ab}$ in the fundamental representation and $N\delta^{ab}$ in the adjoint representation. Moreover, the fermion in the latter case is Majorana, which gives the additional $\frac{1}{2}$ factor compared to the Dirac fermion. Thus, in the latter case we get $\frac{1}{2}N\delta^{ab}$. Their ratio is $1/N \to 0$ at $N \to \infty$.

Problem 6 (L3)

Consider quantal one-dimensional problem described by the Lagrangian

$$\mathcal{L} = \frac{1}{2}\dot{X}^2 - V(X) \tag{B.15}$$

where

$$V(X) = \lambda\left(X^2 - \eta^2\right)^2, \quad \eta \text{ is real and positive} \tag{B.16}$$

(the so-called double-well potential). Find the classical trajectory in the *Euclidean time* connecting the point $X = -\eta$ in the distant past with the point $X = \eta$ in the distant future (keeping in mind that T will be set to ∞ at the end).

Why do we need Euclidean time? We will use this trajectory in subsequent lectures.

Solution

The classical trajectory is determined from the classical equation of motion. If you draw the potential (B.16) you will immediately see that there exists no classical trajectory connecting the points $X = \pm\eta$ at zero energy (we will see shortly that this is the energy we need). However, such a trajectory emerges upon passing to the Euclidean time, $t \to -i\tau$. Indeed, after this transition the Euclidean Lagrangian takes the form

$$\mathcal{L}_{\rm E} = \frac{1}{2}\dot{X}^2 + V(X)\,, \tag{B.17}$$

where now the overdot stands for the derivative over τ, cf. Eq. (3.10). Now, we can interpret the Lagrangian (B.17) as the classical Lagrangian describing the τ-trajectory of the "particle" in the potential $-V$, see Figs. B.2 and B.3. At $\tau \to -\infty$ the particle resides on the top of the hump on the left. If we give it an infinitesimally small velocity directed to the right, at $\tau \to \infty$ it will find itself on the top of the hump on the right. Since both, the potential and kinetic energies at $\tau \to \pm\infty$ vanish and the energy is conserved, we can say that this particular trajectory corresponds to zero energy. If the velocity is finite and positive at $\tau \to \pm\infty$, and so is the kinetic energy, the boundary conditions cannot be satisfied.

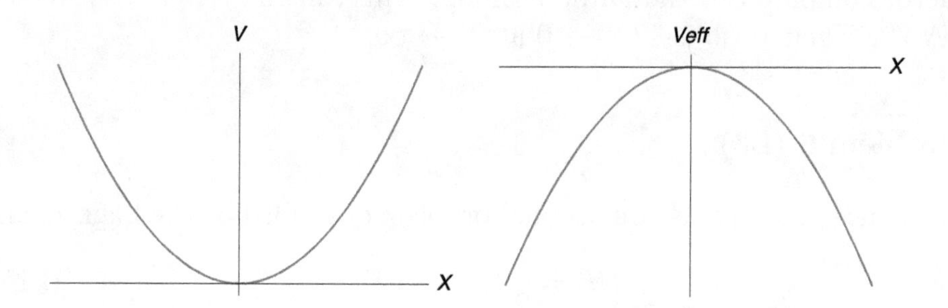

Fig. B.2 Potential for the harmonic oscillator (see Eq. (3.8)) and its effective potential in Euclidean.

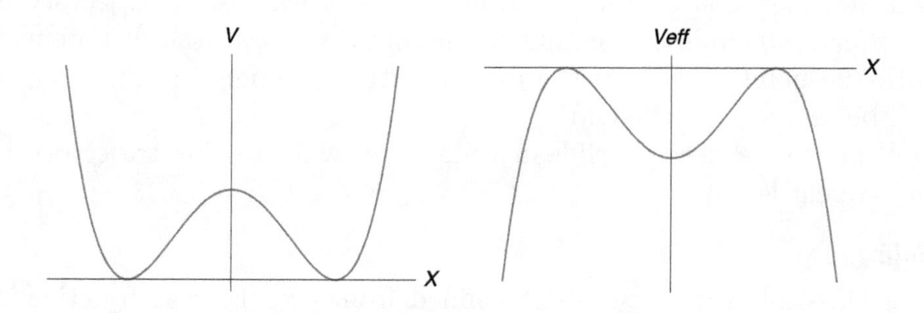

Fig. B.3 Double-well potential (B.16) and its effective potential in Euclidean.

Thus on the classical trajectory in the Euclidean time

$$\frac{1}{2}\dot{X}^2 - V(X) = 0\,, \tag{B.18}$$

which implies

$$\dot{X} = \pm\sqrt{2}\sqrt{V(X)} = \pm\sqrt{2\lambda}(\eta^2 - X^2) \tag{B.19}$$

where the sign choice depends on the boundary conditions. In our case it is +. The solution of the above equation is

$$X_{\rm cl} = \eta \tanh\frac{\omega(\tau - \tau_0)}{2} \tag{B.20}$$

where the parameter ω is conveniently defined as

$$8\lambda\eta^2 = \omega^2\,. \tag{B.21}$$

Explain the meaning of the "extra" arbitrary parameter τ_0 in the solution (B.20).

Problem 7 (L4)

Let us consider

$$\mathcal{L}_J = -\frac{1}{2}\phi(x)\partial^2\phi(x) - \frac{1}{2}m^2\phi^2(x) - \frac{\lambda}{4}\phi^4 + J(x)\phi(x)\,, \tag{B.22}$$

with the real scalar field ϕ and real positive parameters m and λ (the coupling constant λ is assumed to be small). Prove that

$$\left.\frac{\delta^n \log Z[J]}{\delta J(x_1), ...\delta J(x_n)}\right|_{J=0} \equiv 0 \tag{B.23}$$

for all odd values of n.

Solution

At $J = 0$ the Lagrangian given above has Z_2 symmetry under the $\phi \to -\phi$ transformation. If m^2 is positive this symmetry is *not* spontaneously broken. It remains the valid symmetry for all correlation functions.

The expression in (B.23) is

$$\int \mathcal{D}\phi(x)\left(\phi(x_1)...\,\phi(x_n)\right)\,\exp\left\{iS[\phi]\right\}.$$

The pre-exponent is odd under $\phi \to -\phi$ if n is odd resulting in the vanishing of the corresponding n-point function.

Problem 8 (L5)

For a field theory at finite temperature define the free energy F, entropy S, and average energy E in terms of the thermal partition function, Eqs. (5.24) and (5.25).

From this definition find the relation between these three quantities.

Solution

The definition of the (thermal) average energy is obvious,

$$\overline{E} = -\frac{\partial \log Z}{\partial \beta}. \tag{B.24}$$

The free energy F is defined as

$$F = -\frac{\log Z}{\beta}. \tag{B.25}$$

The standard definition of entropy S is

$$S = -\partial F/\partial T$$

which in our notation becomes

$$S = \beta^2 \frac{\partial F}{\partial \beta} = \log Z - \beta \frac{\partial \log Z}{\partial \beta}. \tag{B.26}$$

Equation (B.26) implies in turn that

$$F = \overline{E} - \frac{1}{\beta} S = \overline{E} - T\, S. \tag{B.27}$$

Problem 9 (L5)

Problem 9a: Consider the harmonic oscillator problem discussed in Chapter 3 at finite temperature T, i.e. on S^1 with circumference $\beta = 1/T$ and Lagrangian

$$\mathcal{L} = \frac{\dot{x}^2}{2} - \frac{\omega^2}{2} x^2. \tag{B.28}$$

Impose the periodic boundary conditions and calculate the thermal partition function Z. What would you get with the anti-periodic boundary conditions?

Solution

Solution of this problem can be found in Chapter 3, see the last bracket in Eq. (3.28). For the anti-periodic boundary condition, $\sinh \omega T$ will be replaced by $\cosh \omega T$ leading to "wrong" signs in the partition function.

Problem 9b: Consider the Hamiltonian formulation of the same harmonic oscillator at finite temperature T. Calculate free energy and average energy.

Solution

The Hamiltonian can be written as

$$H = \frac{p^2}{2} + \frac{\omega^2}{2} x^2 \,. \tag{B.29}$$

The energy eigenvalues are

$$E_n = \left(\frac{1}{2} + n\right)\omega, \quad n = 0, 1, 2, ... \tag{B.30}$$

Hence, the partition function is

$$Z(\beta) = \frac{1}{2}\left[\sinh\left(\frac{\beta\omega}{2}\right)\right]^{-1} . \tag{B.31}$$

The free energy is

$$F = -\frac{1}{\beta}\log Z = \frac{\omega}{2} + \frac{1}{\beta}\log\left(1 - e^{-\beta\omega}\right) . \tag{B.32}$$

The average energy is

$$\overline{E} = -\frac{\partial \log Z}{\partial \beta} = \frac{\omega}{2} + \frac{\omega\, e^{-\beta\omega}}{1 - e^{-\beta\omega}} . \tag{B.33}$$

Finally, for the entropy we get

$$S = \log Z - \beta\frac{\partial \log Z}{\partial \beta} = -\log\left(1 - e^{-\beta\omega}\right) + \beta\omega\,\frac{e^{-\beta\omega}}{1 - e^{-\beta\omega}} . \tag{B.34}$$

Let us check Eq. (B.27). To this end we observe that

$$-\frac{S}{\beta} = \frac{1}{\beta}\log\left(1 - e^{-\beta\omega}\right) - \omega\,\frac{e^{-\beta\omega}}{1 - e^{-\beta\omega}} . \tag{B.35}$$

Adding (B.33) and (B.35) we arrive at (B.32).

Problem 10 (L6)

The standard definition of the regular (c-numeric) delta function is

$$\int dx\, \delta(x - y) f(x) \equiv f(y)$$

where $f(x)$ is any continuous function. Define the Grassmann number analog of the delta function. Show that the Grassmann delta function can be represented as a polynomial in the Grassmann variables (i.e. find this polynomial in two cases: $\int d\theta$ and $\int d\theta d\bar{\theta}$ instead of $\int dx$ in the expression above).

Problem 11 (L6)

The Dirac matrices in the spinor representation which I introduced in (6.56) can be defined in many different ways (corresponding to different compositions of the Dirac bispinor from two Weyl spinors), and all of them are unitary equivalent. Compare the definition we used above with those used in Peskin-Schroeder and in Landau-Lifshitz's *Quantum Electrodynamics*. Calculate the matrix

$$\gamma^5 = i \prod_{\mu=0,1,2,3} \gamma^\mu$$

and compare it too. Compare the definitions of left-handed and right-handed. As a reference point use the Dirac equation of motion

$$i\partial_\mu \gamma^\mu \Psi = m\Psi$$

and the fact that neutrino is left-handed. Check that in all definitions of the Dirac matrices the Clifford algebra (6.57) is satisfied.

Warning: Please, remember that raising and lowering the index μ of the Dirac matrices changes the sign of the three spatial matrices. It is standard to compare matrices with the upper vectorial index.

Problem 12* (L6)

Find the unitary transformations connecting two distinct choices of the Dirac matrices.

Solutions to Problems 11 and 12

Problem 11

In Peskin's book,

$$\psi_D = \begin{pmatrix} \psi_L \\ \psi_R \end{pmatrix}. \tag{B.36}$$

In my definition,

$$\psi_D = \begin{pmatrix} \xi_\alpha \\ \bar{\eta}^{\dot{\alpha}} \end{pmatrix}. \tag{B.37}$$

In both Peskin's and my book,

$$\gamma^\mu = \begin{pmatrix} 0 & \sigma^\mu \\ \bar{\sigma}^\mu & 0 \end{pmatrix}, \tag{B.38}$$

and

$$\gamma^5 = i \prod_{\mu=0,1,2,3} \gamma^\mu = \begin{pmatrix} -1 & 0 \\ 0 & 1 \end{pmatrix}, \tag{B.39}$$

Moreover,

$$\{\gamma^\mu, \gamma^\nu\} = \begin{pmatrix} \sigma^\mu \bar{\sigma}^\nu + \sigma^\nu \bar{\sigma}^\mu & 0 \\ 0 & \bar{\sigma}^\mu \sigma^\nu + \bar{\sigma}^\nu \sigma^\mu \end{pmatrix}$$

$$= \begin{cases} 2 & \mu = \nu = 0 \\ 0 & \mu = i, \nu = 0 \text{ or } \nu = i, \mu = 0 \\ \begin{bmatrix} -(\sigma^i\sigma^j + \sigma^j\sigma^i) & 0 \\ 0 & -(\sigma^i\sigma^j + \sigma^j\sigma^i) \end{bmatrix} & \mu = i, \nu = j \end{cases}$$

$$= \begin{cases} 2 & \mu = \nu = 0 \\ 0 & \mu = i, \nu = 0 \text{ or } \nu = i, \mu = 0 \\ -2\delta^{ij} & \mu = i, \nu = j \end{cases} \tag{B.40}$$

Thus, indeed

$$\{\gamma^\mu, \gamma^\nu\} = 2g^{\mu\nu},$$

so the Clifford algebra is satisfied. In Landau's book,

$$\psi_D = \begin{pmatrix} \xi^\alpha \\ \eta_{\dot{\alpha}} \end{pmatrix}, \tag{B.41}$$

$$\gamma^\mu = \begin{pmatrix} 0 & \bar{\sigma}^\mu \\ \sigma^\mu & 0 \end{pmatrix}. \tag{B.42}$$

Moreover,

$$\gamma^5 = i \prod_{\mu=0,1,2,3} \gamma^\mu = \begin{pmatrix} 1 & 0 \\ 0 & -1 \end{pmatrix}. \tag{B.43}$$

We can check the Clifford algebra too,

$$\{\gamma^\mu, \gamma^\nu\} = \begin{pmatrix} \bar{\sigma}^\mu \sigma^\nu + \bar{\sigma}^\nu \sigma^\mu & 0 \\ 0 & \sigma^\mu \bar{\sigma}^\nu + \sigma^\nu \bar{\sigma}^\mu \end{pmatrix}$$

$$= \begin{cases} 2 & \mu = \nu = 0 \\ 0 & \mu = i, \nu = 0 \text{ or } \nu = i, \mu = 0 \\ \begin{bmatrix} -(\sigma^i \sigma^j + \sigma^j \sigma^i) & \\ & -(\sigma^i \sigma^j + \sigma^j \sigma^i) \end{bmatrix} & \mu = i, \nu = j \end{cases}$$

$$= \begin{cases} 2 & \mu = \nu = 0 \\ 0 & \mu = i, \nu = 0 \text{ or } \nu = i, \mu = 0 \\ -2\delta^{ij} & \mu = i, \nu = j \end{cases} \tag{B.44}$$

$$= 2g^{\mu\nu} . \tag{B.45}$$

So the Clifford algebra is satisfied.

The Dirac equation for neutrino in different representations:

Peskin's and my book

$$i\bar{\sigma}^\mu \partial_\mu \xi_\alpha = 0,$$

Landau's book

$$i\sigma^\mu \partial_\mu \xi^\alpha = 0 . \tag{B.46}$$

The difference is because the two equations are written in two different representations of γ matrices, corresponding to two distinct ξ.

Problem 12

$$\gamma_1^\mu = \begin{pmatrix} 0 & \sigma^\mu \\ \bar{\sigma}^\mu & 0 \end{pmatrix},$$

$$\gamma_2^\mu = \begin{pmatrix} 0 & \bar{\sigma}^\mu \\ \sigma^\mu & 0 \end{pmatrix}. \tag{B.47}$$

Therefore the unitary transformation U is

$$U = \begin{pmatrix} 0 & 1_{2\times 2} \\ 1_{2\times 2} & 0 \end{pmatrix}, \tag{B.48}$$

so that,

$$U^\dagger \gamma_1^\mu U = \gamma_2^\mu, \text{ for all } \mu . \tag{B.49}$$

Problem 13 (L7)

Assume that you have a Yang-Mills theory with the gauge group

$$SU(N) \times SU(N)\,,$$

with a single Dirac fermion field which transforms in the fundamental representation of the first $SU(N)$ and *anti*fundamental representation of the second $SU(N)$. Write the expression for the covariant derivative for this fermion field. Assume that the gauge coupling constants for both $SU(N)$'s are equal. What additional discrete global symmetry does this theory acquire in this case?

Problem 14* (L7)

As was mentioned in the title in the very beginning of this Chapter, all equations above refer to the Dirac fermions. What problems do you see if you try to generalize the above derivations to the Weyl or Majorana fermions?

Assume you consider two distinct Yang-Mills theories, with fermions in a real representation of the gauge group (e.g. adjoint). In the first theory the fermions are Dirac, in the second Majorana. Find the relation between the determinants

$$\det\,(iD_\mu\gamma^\mu - m)_{\text{Dirac}} \text{ and } \det\,(iD_\mu\gamma^\mu - m)_{\text{Majorana}}\,.$$

Solution

$$\det\,(iD_\mu\gamma^\mu - m)_{\text{Majorana}} = \sqrt{\det\,(iD_\mu\gamma^\mu - m)_{\text{Dirac}}}\,.$$

Hint: Every Dirac spinor can be represented as a linear combination of two independent Majorana spinors. For the adjoint fermions the two corresponding terms for the two Majorana fields decouple from each other.

Problem 15 (L8)

In QFT-I, we discussed the photon propagator $G(k)$ in QED in the so-called Coulomb gauge, in which $\partial_i A_i = 0$ (or in the momentum space

$k_i A_i = \vec{k}\vec{A} = 0$). In QFT-I, we relied on the canonic quantization rather than the path-integral quantization. In some textbooks it is stated that in the Coulomb gauge one can set $A_0 = 0$. If so, it would seemingly imply that the $G^{00} = 0$ where G^{00} is the 00 component of the Green's function,

$$G^{\mu\nu}(x-y) = \langle 0|T\Big(A^\mu(x)A^\nu(y)\Big)|0\rangle\,. \tag{B.50}$$

Explain why (i) A^0 cannot be set to zero in the theory with electrons as opposed to pure photodynamics (no interactions!); and (ii) as a result, $G^{00} \neq 0$ in the Coulomb gauge.

Find the relation between the photon propagator in the Feynman gauge

$$G^{\mu\nu}(k) = -i\,\frac{g^{\mu\nu}}{k^2} \tag{B.51}$$

and that in the Coulomb gauge, in the form

$$G^{\mu\nu}_{\rm Coulomb}(k) = (-i)\,\frac{1}{k^2}\Big(g^{\mu\nu} - k^\mu B^\nu - k^\nu B^\mu\Big). \tag{B.52}$$

Explain your result for the four-vector B^μ and G^{00}.

Solution

One can start from the classical equation of motion

$$\partial^2 A^\nu(x) - \partial^\nu(\partial_\mu A^\mu(x)) = j^\nu(x)\,, \tag{B.53}$$

where $j^\nu(x)$ is the electron (positron) source. Formally, inverting this equation we can write

$$A^\mu(x) = -i\int d^4y\; G^{\mu\nu}(x,y)\, j_\nu(y)\,. \tag{B.54}$$

Since in the Coulomb gauge $\vec{\partial}\vec{A} = 0$, we can write that $\partial_\mu A^\mu(x) = \partial_0 A^0$.

Then the time component of Eq. (B.53) takes the form

$$-\,\vec{\partial}^2 A^0 = j^0\,. \tag{B.55}$$

The time derivative drops out! Hence, the inverse of Eq. (B.55) takes the form

$$A^0(t,\vec{x}) = -\vec{\partial}^{-2} j^0 = \int d^3\vec{y}\;\frac{1}{4\pi|\vec{x}-\vec{y}|}\; j^0(t,\vec{y})\,. \tag{B.56}$$

This is the instantaneous Coulomb interaction.

As a result, comparing with Eq. (B.54), we obtain in the coordinate space

$$G^{00}(x,y) = i\,\delta(x^0 - y^0)\,\frac{1}{4\pi|\vec{x}-\vec{y}|} \to \frac{i}{\vec{k}^2}\,. \qquad \text{(B.57)}$$

The arrow on the right-hand side of the above equation means transition to the momentum space. Using Eq. (B.57) and the conditions $G^{0i} = 0$ and $k_i G^{ij} = 0$, we arrive at

$$G^{\mu\nu}_{\text{Coulomb}}(k) = \frac{-i}{k^2}\left(g^{\mu\nu} - k^\mu B^\nu - k^\nu B^\mu\right) \qquad \text{(B.58)}$$

where

$$B^\mu = \frac{(k^0, -\vec{k})}{2\vec{k}^2}\,. \qquad \text{(B.59)}$$

Equation (B.59) allows one to reproduce (B.57) and $G^{0i} = 0$, while for G^{ij} we get

$$G^{ij}_{\text{Coulomb}}(k) = \frac{i}{k^2}\left(\delta^{ij} - \frac{k^i k^j}{\vec{k}^2}\right). \qquad \text{(B.60)}$$

Problem 16 (L8-L12)

Generalize Eqs. (8.23), and (8.24)–(8.26) to non-Abelian case. For each given value of the gauge parameter ξ (see the end of the previous lecture) find the kinetic term for the gauge bosons, i.e. the term in the Lagrangian (after integrating out $\omega(x)$) quadratic in the fields A^a_μ and quadratic in derivatives. Write this term in the momentum space, inverse the kinetic energy-momentum operator in the Lagrangian and find the propagator as a function of ξ. Find the values of ξ corresponding to the Landau and Feynman gauges.

Solution

The gluon propagator defined as

$$G^{ab}_{\mu\nu} = \langle 0|T\Big(A^a_\mu(k)\;A^b_\nu(-k)\Big)|0\rangle \qquad \text{(B.61)}$$

has the form

$$G^{ab}_{\mu\nu} = -\frac{i}{k^2}\left[\left(g_{\mu\nu} - \frac{k_\mu k_\nu}{k^2}\right) + \xi\frac{k_\mu k_\nu}{k^2}\right]\delta^{ab}\,. \qquad \text{(B.62)}$$

In the Feynman gauge $\xi = 1$ while in the Landau gauge $\xi = 0$. The Landau gauge is sometimes referred to as the Lorentz gauge.

Problem 17 (L8-L12)

Assume that you have a Yang-Mills theory with the gauge group

$$SU(3) \times SU(2) \times U(1)\,,$$

as in the standard model. Do the path-integral gauge fixing in this case (assuming that you know how to do it for each separate gauge factor, which – I hope – you do). How many ξ parameters do you need? What is the number of independent gauge coupling constants?

Solution

For the $SU(3) \times SU(2) \times U(1)$ gauge group we introduce three distinct gauge potentials $A_\mu^{(i)a_i}$ where $i = 1, 2, 3$ mark gauge factors in the product group (i coincides with the rank of the given gauge factor), namely, $a_1 = 1$ for $U(1)$, $a_2 = 1, 2, 3$ for $SU(2)$, and $a_3 = 1, 2, ..., 8$ for $SU(3)$. The action is

$$S = \sum_i S_i \tag{B.63}$$

where

$$S_i = \int d^4x \, \frac{1}{g_i^2} \left\{ -\frac{1}{4} F_{\mu\nu}^{a_i} F^{\mu\nu\, a_i} - \frac{1}{2\xi_i} \left(\partial^\mu A_\mu^{(i)a_i} \right)^2 + \partial_\mu \bar{c}_i^{a_i} D^\mu c_i^{a_i} \right\}. \tag{B.64}$$

No summation over i is to be performed in (B.64). The covariant derivatives and $F_{\mu\nu}^{a_i}$ are defined using the i-th gauge potential. Moreover, $\bar{c}_i$ and c_i represent three sets of ghosts in the adjoint representation for each factor group. For the $U(1)$ factor group the ghost term can be dropped because it does not depend on the gauge potential.

The action above depends on three independent gauge parameters ξ_i and three gauge couplings g_i.

Problem 18 (L10)

Read pp. 249-250 in Peskin and Schroeder on dimensional regularization (DR). Calculate (10.13) using dimensional regularization. Compare (10.16) and the DR result. Establish the correspondence between $\log M_0^2$ in the PV regularization and $1/\varepsilon$ in the DR method.

Solution

$$\log M_0^2 = \frac{2}{\varepsilon} + \log 4\pi - \gamma$$

where γ is the Euler constant.

Problem 19 (L12)

Complete the derivation in Sec. 12.3, page 115, taking into account the h field.

Problem 20 (L13)

In the O(3) sigma model

$$\mathcal{L} = \frac{1}{2g^2} \left(\partial_\mu \vec{S}\right) \left(\partial^\mu \vec{S}\right) \tag{B.65}$$

use the constraint $\vec{S}^2 = 1$ to express $\vec{S}_3$ in terms of $\vec{S}_{1,2}$ and then find the Lagrangian (B.65) in terms of the unconstraint components $\vec{S}_{1,2}$. The first two terms determine the propagators of the $S_{1,2}$ fields, the third term represents interactions between them.

Solution

$$S_3 = \sqrt{1 - S_1^2 - S^2}, \qquad \partial S_3 = -\frac{S_1 \partial S_1 + S_2 \partial S_2}{\sqrt{1 - S_1^2 - S^2}}\,. \tag{B.66}$$

The latter expression implies

$$\mathcal{L} = \frac{1}{2g^2}\left[(\partial S_1)^2 + (\partial S_2)^2 + \frac{(S_1 \partial S_1 + S_2 \partial S_2)^2}{1 - S_1^2 - S^2}\right]. \tag{B.67}$$

Problem 21 (L14)

Consider the Standard Model with three generations. Using general formulas that we already know from the previous lectures find the first coefficients in the β functions in the SU(2) and U(1) sectors of this model. Assume that $\mu \gg gv$ where v is the vacuum expectation value of the Higgs field.

Solution

The first coefficient of the β function for the $SU(2)$ sector is

$$b_{SU(2)} = \frac{22}{3} - \frac{1}{3} N_{\text{gen}} (1 + N_{\text{color}}) - \frac{1}{6}, \tag{B.68}$$

where N_{gen} is the number of generation, $N_{\text{gen}} = 3$ in the Standard Model. The contribution proportional to N_{gen} comes from fermions while the last term is due to the Higgs field loop. The factor $\frac{1}{3}$ instead of $\frac{2}{3}$ in front of N_{gen} appears due to the fact that only the left-handed fermions are charged with respect to $SU(2)$. The contribution proportional to the number of colors N_{color} comes from quarks ($N_{\text{color}} = 3$).

For the $U(1)$ β function we obtain

$$b_{U(1)} = -\frac{2}{3} N_{\text{gen}}\, c - \frac{1}{3} c_H, \tag{B.69}$$

where the last contribution is due to the Higgs field. The constant c is

$$c = \frac{1}{4} \left\{ \sum_l \left(e_l^L\right)^2 + \sum_l \left(e_l^R\right)^2 + N_{\text{color}} \left[\sum_q \left(e_q^L\right)^2 + \sum_q \left(e_q^R\right)^2 \right] \right\}, \tag{B.70}$$

where the sums run over the leptons (index l) and quarks (index q) for each generation, and the couplings $e_{l,q}^{L,R}$ correspond to weak hypercharges for the left- and right-handed fermions. They are

$$e_l^L = (-1, -1), \quad e_l^R = (2, 0), \quad \text{electron and neutrino, respectively,} \tag{B.71}$$

and

$$e_q^L = \left(\frac{1}{3}, \frac{1}{3}\right), \quad e_q^R = \left(-\frac{4}{3}, \frac{2}{3}\right), \quad u \text{ and } d \text{ quarks, respectively}. \tag{B.72}$$

The same charge assignment applies to other generations. As for the constant c_H in the second term of (B.69) we find

$$c_H = \frac{2}{4} = \frac{1}{2}, \tag{B.73}$$

where the factor 2 comes from the Higgs *doublet* with the unit weak hypercharge.

Problem 22 (L14)

Consider the Standard Model with three generations, as in Problem 20, assuming, however, the opposite limit $\mu \ll gv$ where v is the vacuum expectation value of the Higgs field and μ is the normalization point in the RG flow. Assume also that $\mu \gg m_{q,l}$ where $m_{q,l}$ are the lepton and quark masses. For educational purposes in this exercise, let us close our eyes to the fact that the t quark is too heavy for this inequality to be fully valid.

Using the general formulas we know from the previous lectures, determine the first coefficients in the β function in this model (remember, at $\mu \ll v$ the only gauge bosons which survives is the photon).

Solution

At energies below v we deal with QED. The first coefficient of the photon β function is

$$b_\gamma = -\frac{2}{3} N_{\rm gen}\, c\,, \tag{B.74}$$

where

$$\begin{aligned} c &= \frac{1}{4}\left\{\sum_l \left(e_l^L\right)^2 + \sum_l \left(e_l^R\right)^2 + N_{\rm color}\left[\sum_q \left(e_q^L\right)^2 + \sum_q \left(e_q^R\right)^2\right]\right\} \\ &= 2\left\{\sum_l \left(e_l^L\right)^2 + N_{\rm color}\sum_q \left(e_q^L\right)^2\right\}. \end{aligned} \tag{B.75}$$

Here $N_{\rm color} = 3$, and the sums run over the leptons (index l) and the quarks (index q) for each generation. Moreover, unlike in Problem 20, $e_{l,q}^{L,R}$ are the *electric* charges for the left- and right-handed fermions,

$$\begin{aligned} &e_l^L = (-1, 0)\,, \quad e_l^R = (1, 0)\,, \quad \text{for electron and neutrino;} \\ &e_q^L = \left(\frac{2}{3}, -\frac{1}{3}\right), \quad e_q^R = \left(-\frac{2}{3}, \frac{1}{3}\right), \quad \text{for } u \text{ and } d \text{ quarks.} \end{aligned} \tag{B.76}$$

The electric charge assignments for other generations are the same.

Problem 23

Assume we have Yang-Mills theory with the gauge sector

$$G = SU(2)\,,$$

and the Higgs sector consisting of one real scalar field in the *adjoint* representation of $SU(2)$.

The adjoint field develops a vacuum expectation value $\langle \phi^a \phi^a \rangle = v^2$. Calculate the first coefficients in the β function for $\mu \gg gv$. Compare the result to the first coefficient in the β function of pure Yang-Mills theory with the same gauge group as above. Can you give a physical interpretation for the difference?

Solution

The first coefficient of the β function in the given problem is

$$b = \frac{22}{3} - \frac{1}{3} = 7\,, \qquad \mu \gg gv\,. \tag{B.77}$$

This has to be compared with $\frac{22}{3}$ representing the contribution of the uh-Higgsed gauge bosons. The term $-\frac{1}{3}$ in (B.77) can be explained by longitudinal components of the Higgsed gauge bosons.

Problem 24

Assume that you have Yang-Mills theory with the gauge group $SU(N)$ and a single *Majorana* fermion in the *adjoint* representation. Calculate its contribution to the gauge coupling renormalization.

Why cannot the fermion field in the fundamental representation of $SU(N)$ be Majorana fermion?

Solution

The first coefficient of the *beta* function is

$$b = \frac{11}{3}N - \frac{2}{3}N = 3N\,, \tag{B.78}$$

where $-\frac{2}{3}$ instead of $-\frac{4}{3}$ takes into account the Majorana nature of the fermion.

The fundamental representation in $SU(N)$ with $N > 2$ is complex. The Majorana spinor can be defined as real.

In fact, the theory presented in this problem is minimally supersymmetric Yang-Mills theory.

Problem 25 (L16)

Calculate the anomalous dimensions for the following operators (in the framework of QCD, at one loop):

$$\bar{\psi}\gamma^5\psi\,, \quad \bar{\psi}\gamma^\mu\psi\,, \quad \bar{\psi}\gamma^\mu\gamma^5\psi\,, \quad \bar{\psi}\sigma^{\mu\nu}\psi\,. \tag{B.79}$$

It is implied that the normalization point μ is much higher then the quark mass m so the latter can be neglected.

Solution

The required Z factors are given by the diagram of Fig. 16.1(b) where the closed circle in the given problem denotes the insertion of one of the operators $\mathcal{O}_1 = \bar{\psi}\gamma^5\psi$, $\mathcal{O}_2 = \bar{\psi}\gamma^\mu\psi$, $\mathcal{O}_3 = \bar{\psi}\gamma^\mu\gamma^5\psi$, and $\mathcal{O}_3 = \bar{\psi}\sigma^{\mu\nu}\psi$, respectively, rather than the mass insertion considered in Chapter 16. The calculation of the graphs in Fig. 16.1(b) produces the Z factors analogous to (16.9),

$$\begin{aligned}
Z_{\bar{\psi}\gamma^5\psi} &= 1 + 4\frac{N^2-1}{2N}\frac{g^2}{16\pi^2}\log\frac{M^2}{\mu^2}\,, \\
Z_{\bar{\psi}\gamma^\mu\psi} &= 1 + \frac{N^2-1}{2N}\frac{g^2}{16\pi^2}\log\frac{M^2}{\mu^2}\,, \\
Z_{\bar{\psi}\gamma^\mu\gamma^5\psi} &= 1 + \frac{N^2-1}{2N}\frac{g^2}{16\pi^2}\log\frac{M^2}{\mu^2}\,, \\
Z_{\bar{\psi}\sigma^{\mu\nu}\psi} &= 1\,.
\end{aligned} \tag{B.80}$$

As in Chapter 16 we use the Feynman gauge. For $N = 3$ the above relations take the form

$$\begin{aligned}
Z_{\bar{\psi}\gamma^5\psi} &= 1 + 4\frac{4}{3}\frac{g^2}{16\pi^2}\log\frac{M^2}{\mu^2}\,, \\
Z_{\bar{\psi}\gamma^\mu\psi} &= 1 + \frac{4}{3}\frac{g^2}{16\pi^2}\log\frac{M^2}{\mu^2}\,, \\
Z_{\bar{\psi}\gamma^\mu\gamma^5\psi} &= 1 + \frac{4}{3}\frac{g^2}{16\pi^2}\log\frac{M^2}{\mu^2}\,, \\
Z_{\bar{\psi}\sigma^{\mu\nu}\psi} &= 1\,.
\end{aligned} \tag{B.81}$$

The unity in the last line of (B.81) is a trivial consequence of the trace of the appropriate γ matrices.

The anomalous dimensions of the operators (B.79) are obtained by dividing the above Z factors by Z_Ψ presented in Eq. (16.8), which yields

$$1+4L\,,\quad 1\,,\quad 1\,,\quad 1-\frac{4}{3}L\,, \tag{B.82}$$

respectively, where

$$L \equiv \frac{g^2}{16\pi^2}\log\frac{M^2}{\mu^2}\,. \tag{B.83}$$

I recommend to repeat the above calculations in the Landau gauge in which the numerator of the gluon Green function is proportional to $g^{\mu\nu} - q^\mu q^\nu/q^2$. Hint: in this gauge $Z_\Psi = 1$ (i.e. the coefficient of logarithm (B.83) vanishes).

Problem 26* (L17)

The unitary matrices spanning $SU(2)$ used in the effective chiral Lagrangian (17.9) are defined in Eq. (17.7) which presents a standard parametrization of the pion fields. Equation (17.7) is just one possible parametrization. Another (often used) parametrization of the $SU(2)$ matrices is

$$U = \left(1 - \frac{i\tau^a\varphi^a}{2F_\pi}\right)\left(1 + \frac{i\tau^a\varphi^a}{2F_\pi}\right)^{-1}$$

$$U^\dagger = \left(1 - \frac{i\tau^a\varphi^a}{2F_\pi}\right)^{-1}\left(1 + \frac{i\tau^a\varphi^a}{2F_\pi}\right). \tag{B.84}$$

It is obvious that $U^\dagger U = 1$. The above parametrization is alternative to (17.7) and can be used in (17.9). Find the corresponding expression for the effective Lagrangian and the relation between the fields π^a and φ^a.

Solution

The analog of the Lagrangian (17.9) is

$$\mathcal{L} = \frac{F_\pi^2}{4}\mathrm{Tr}\left(\partial_\mu U \partial^\mu U^\dagger\right) \tag{B.85}$$

where U, $U^\dagger$ are defined in (B.84). Note that the matrix U can be conveniently rewritten as

$$U = \left(1 - \frac{\vec{\varphi}^2}{4F_\pi^2} + \frac{i\vec{\tau}\vec{\varphi}}{F_\pi}\right)\left(1 + \frac{\vec{\varphi}^2}{4F_\pi^2}\right)^{-1}. \tag{B.86}$$

Substituting (B.86) in (B.85) we arrive at

$$\mathcal{L} = \frac{1}{2}\,\frac{\partial_\mu \varphi^a \partial^\mu \varphi^a}{\left(1 + \frac{\varphi^a \varphi^a}{4F_\pi^2}\right)^2}\,, \tag{B.87}$$

Moreover, comparing (17.7) and (B.87) we conclude that

$$\begin{aligned} \pi^a &= \varphi^a \mathcal{F}(y)\,, \\ \mathcal{F}(y) &= \frac{1}{2\sqrt{y}}\,\text{Arccos}\frac{1-y}{1+y} = 1 - \frac{y}{3} + \frac{y^2}{5} - \frac{y^3}{7} + \ldots\,, \\ y &\equiv \frac{\vec{\varphi}^{\,2}}{4F_\pi^2}\,. \end{aligned} \tag{B.88}$$

The effective chiral Lagrangian (17.9) in the parametrization (17.7) takes the form

$$\mathcal{L} = \frac{1}{2}\left[\partial_\mu \vec{\pi}\,\partial^\mu \vec{\pi}\left(\frac{1}{\pi^2}\,\sin^2\pi\right) + \partial_\mu \pi\,\partial^\mu \pi\left(1 - \frac{1}{\pi^2}\,\sin^2\pi\right)\right] \tag{B.89}$$

where

$$\pi \equiv \sqrt{\vec{\pi}\vec{\pi}} = \sqrt{\pi^a \pi^a}\,, \qquad a = 1, 2, 3. \tag{B.90}$$

Verification: Start from (17.9) and expand this formula up to the fourth order in π. Then expand Eq. (B.87) up to the fourth order in φ. Use Eq. (B.88) to demonstrate that these expansions are consistent with each other up to this order.

Generalities: In the general case (i.e. for an arbitrary parametrization of the target space) any chiral Lagrangian can be written in the form

$$\mathcal{L} = \frac{1}{2}\,g_{ab}(\varphi)\,\partial_\mu \pi^a\,\partial^\mu \pi^b\,, \tag{B.91}$$

where $g_{ab}(\varphi)$ is the metric on the target space. For (B.87) we have the simplest metric

$$g_{ab} = \delta_{ab}\frac{1}{\left(1 + \frac{\varphi^a \varphi^a}{4F_\pi^2}\right)^2}\,. \tag{B.92}$$

The same target space in the π parametrization is described by a different metric which can be readily inferred from (B.89).

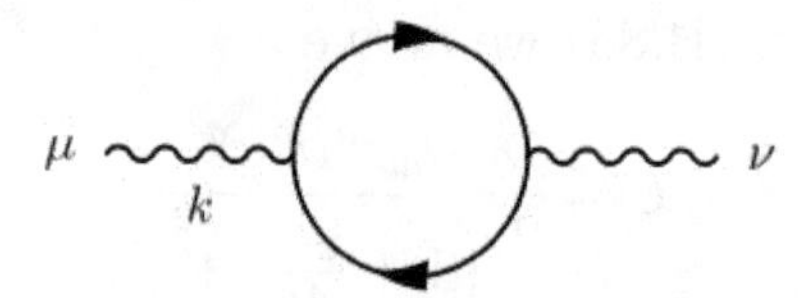

Fig. B.4 Calculation of the one-loop radiative correction to the photon propagator in the Schwinger model.

Problem 27 (L18)

*Part 27a**

Consider the fermion (electron) loop depicted in Fig. B.4 in *two-dimensional* quantum electrodynamics (also known as the Schwinger model). Find the correction due to this Feynman diagram to the tree-level photon propagator. Make sure that in your calculation transversality is preserved. The Lagrangian of the Schwinger model is

$$\mathcal{L} = -\frac{1}{4e^2} F_{\mu\nu}F^{\mu\nu} + \bar{\psi}\,\gamma^\mu\,(i\partial_\mu + A_\mu)\,\psi\,, \tag{B.93}$$

where ψ is the two-dimensional Dirac spinor and e is the electromagnetic coupling constant. The Dirac matrices γ^μ are defined in Eq. (6.66).

Solution

This problem in four dimensions was solved on page 98. I will use the same method of calculation.

The two-dimensional case is somewhat more subtle. A straightforward calculation similar to that carried out on page 98 would require to expand our expressions in ϵ for $D = 2 - \epsilon$. (The reason will become clear shortly.) One can avoid this by observing that

$$\frac{1}{(x^2)^2}\left(2x^\mu\,x^\nu - g^{\mu\nu}x^2\right) \equiv -\partial_\mu\,\frac{x_\nu}{x^2}\,, \tag{B.94}$$

and then performing integration by parts in the Fourier integral. Using Eq. (10.50) we obtain

$$\frac{1}{(x^2)^2}\left(2x^\mu\,x^\nu - g^{\mu\nu}x^2\right) \longrightarrow ip^\mu\,(-2\pi)\frac{p^\nu}{p^2}\,. \tag{B.95}$$

This expression is not transversal. That's because we dropped a contact term which has no p dependence whatsoever. Restoring this contact term from the requirement of transversality we arrive at

$$\Pi^{\mu\nu}(p) = \frac{e^2}{2\pi}\left(-\frac{p^\mu p^\nu}{p^2} + g^{\mu\nu}\right), \qquad D = 2\,. \tag{B.96}$$

Equation (B.96) does not contain logarithms. That's the reason why deriving it from (10.47) and (10.51) would require the expansion in ϵ for $D = 2 - \epsilon$.

Part 27b

Find the anomaly in the divergence of the axial current in the Schwinger model (B.93). Use the Pauli-Villars regularization as in Sec. 18.2.1.

Solution

The axial current regularized by the Pauli-Villars term has the form

$$J_A^5 = \bar\psi\gamma^\mu\gamma^5\psi + \bar R\gamma^\mu\gamma^5 R\,, \tag{B.97}$$

where the γ matrices are defined in (6.66) and R is the regulator field as in Sec. 18.2.1. The mass of the R field $M \to \infty$ at the end of the calculation, so it disappears from the physical spectrum. Its statistic is opposite to that of a normal fermion field, so that the regulator loop does not have the minus sign inherent to normal fermion loops. The only diagram surviving in the limit $M \to \infty$ is diangle rather than triangle emerging in four dimensions. See Fig. B.5. This follows from the mass dimension of the current in (B.97), which is 1 in contradistinction with 3 in four dimensions.

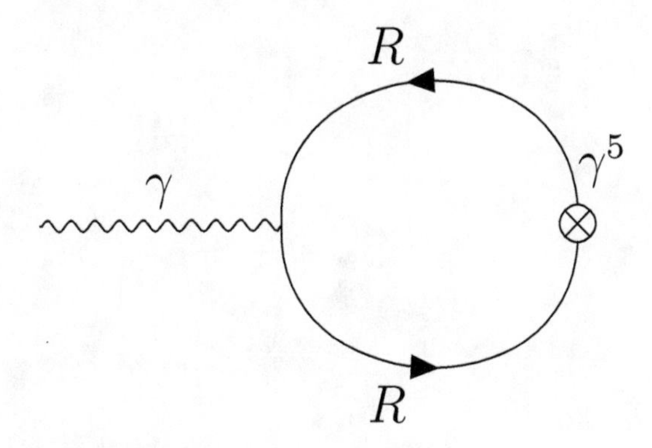

Fig. B.5 Calculation of the anomaly in the divergence of the axial current in the Schwinger model using the Pauli-Villars regularization.

We start from

$$\partial_\mu J_A^5 = 2iM\bar{R}\gamma^5 R \tag{B.98}$$

observing that the classical part of the divergence vanishes, as was expected. Then we insert the vertex $\bar{R}\gamma^5 R$ in the anomalous diangle diagram and obtain

$$\partial_\mu J_A^5 = (2iM)\, iA_\mu \int \frac{d^2p}{4\pi^2} \mathrm{Tr}\left[\gamma^\mu \frac{i(\not{p}+M)}{p^2-M^2}\gamma^5 \frac{i(\not{p}-\not{q}+M)}{(p-q)^2-M^2}\right]. \tag{B.99}$$

Since we are seeking the contribution that is non-vanishing in the $M \to \infty$ limit, significant simplifications can be made. It is not difficult to understand that in the first propagator we can drop $\not{p}$ in the numerator. In the second propagator, we can drop q in the denominator. After this is done, the numerator of the second propagator can also be simplified. Indeed, because of the trace over the spinor indices, the M term drops out. The $\not{p}$ term gives no contribution because of the angular integration.

After rotating to the Euclidean space and carrying out integration we arrive at

$$(2iM)\,\frac{1}{2\pi}\,\varepsilon^{\mu\alpha}\,A_\mu q_\alpha\,\frac{1}{M}\,. \tag{B.100}$$

Equation (B.100) implies, in turn, that

$$\partial_\mu J_A^5 = -\frac{1}{2\pi}\varepsilon^{\mu\alpha}F_{\mu\alpha}\,. \tag{B.101}$$

This is a two-dimensional analogue of Eq. (18.10).

Appendix C

Brief History of QFT

Advent of Quantum Field Theory

Based on S. Weinberg's Essays [1] and Other Sources [1]

It is widely supposed that progress in science occurs in large or small revolutions. In this view, the successes of previous revolutions tend to fasten upon the scientist's mind a language, a mind-set, a body of doctrine, from which he must break free in order to advance further. There is great debate about the degree to which these revolutions are brought about by the individual scientific genius, able to transcend the fixed ideas of his times, or by the accumulation of discrepancies between existing theory and experiment. However, there seems to be a general agreement that the essential element of scientific progress is a decision to break with the past.

I would not quarrel with this view, as applied to many of the major advances in the history of science. It certainly seems to apply to the great revolutions in physics in the 20th century: the development of special relativity and of quantum mechanics. However, the development of quantum field theory since 1930 provides a curious counterexample, in which the essential element of progress has been the realization, again and again, that a revolution is unnecessary. If quantum mechanics and relativity were revolutions in the sense of the French Revolution of 1789 or the Russian Revolution of 1917, then quantum field theory is more of the order of the Glorious Revolution of 1688,[2] things changed only just enough so that they could stay the same.

When quantum mechanics was developed, physicists already knew about various *classical* fields, notably the electromagnetic field. The first application of the new quantum mechanics of 1925-1926 to fields

[1]The first 3.5 pages represent selected quotations from [1].

[2]The Glorious Revolution of 1688 was the overthrow of King James II of England by a union of English Parliamentarians with the Dutch *stadtholder* William III of Orange-Nassau (William of Orange). William's successful invasion of England with a Dutch fleet and army led to him ascending the English throne as William III of England jointly with his wife Mary II of England, in conjunction with the documentation of the Bill of Rights 1689.

rather than particles came in one of the founding papers of quantum mechanics itself. In 1926, Born, Heisenberg, and Jordan [2] turned their attention to the electromagnetic field in empty space, in the absence of any electric charges or currents.

For simplicity, they left out the polarization of the photon, and took spacetime to have one space and one time dimension, but that didn't affect the main results. Born *et al.* gave a formula for the electromagnetic field as a Fourier transform and used the canonical commutation relations to identify the coefficients in this Fourier transform as operators that destroy and create photons, so that when quantized this field theory became a theory of photons. Photons, of course, had been around (though not under that name) since Einstein's work on the photoelectric effect two decades earlier. The word photon was coined in 1926 by the optical physicist Frithiof Wolfers and the chemist Gilbert N. Lewis.

By applying to the electromagnetic field the same mathematical methods that they had used for material oscillators, Born *et al.* were able to show that the energy of each mode of oscillation of an electromagnetic field is quantized – the allowed value, are separated by a basic unit of energy, given by the frequency of the mode times $\hbar$. The physical interpretation of this result was immediate. The state of the lowest energy is the radiation-free empty space, and can be assigned an energy equal to zero. The next lowest state must then have an energy equal to the frequency times $\hbar$, and can be interpreted as the state of a single photon with that energy. The next state would have an energy twice as great, and therefore would be interpreted as containing two photons of the same energy. And so on. Thus, the application of quantum mechanics to the electromagnetic field had at last put Einstein's idea of the photon on a firm mathematical foundation.

Born, Heisenberg, and Jordan had dealt only with the electromagnetic field in empty space, so although their work was illuminating, it did not lead to any important quantitative predictions. The inception of the first "practical" quantum field theory – four-dimensional quantum electrodynamics – is usually attributed to a 1927 paper of Paul Adrian Maurice Dirac. [3]. The first paper in [3] is entitled "The Quantum Theory of the Emission and Absorbtion of Radiation." Namely here Dirac coined the name quantum electrodynamics (QED).

Dirac was grappling with an old problem: how to calculate the rate at which atoms in excited states would emit electromagnetic radiation and drop into states of lower energy. The difficulty was not so much in deriving an answer –the correct formula had already been derived in an *ad hoc* sort of way by Born and Jordan [4] and by Dirac himself [5]. The problem was to understand this guessed-at formula as a mathematical consequence of quantum mechanics. This problem was of crucial importance, because the process of spontaneous emission of radiation is one in which "particles" are actually created. Before the event, the system consists of an excited atom, whereas after the event, it consists of an atom in a state of lower energy, plus one photon. If quantum mechanics could not deal with process of creation and destruction, it could not be an all-embracing physical theory.

Dirac's successful treatment of the spontaneous emission of radiation confirmed the universal character of quantum mechanics. However, the world was still conceived to be composed of two very different ingredients – particles and fields – which were both to be described in terms of quantum mechanics, but in very different ways. Material particles like electrons and protons were conceived to be eternal.

On the other hand, photons were supposed to be merely a manifestation of an underlying entity, the quantized electromagnetic field, and could be freely created and destroyed.

It was not long before a way was found out of this distasteful dualism, toward a truly unified view of nature. The essential steps were taken in a 1928 paper of Jordan and Eugene Wigner [6], and then in a pair of long papers in 1929-1930 by Heisenberg and Pauli [7]. They showed that material particles could be understood as the quanta of various fields, in just the same way that the photon is the quantum of the electromagnetic field. There was supposed to be one field for each type of elementary particle. Thus, the inhabitants of the universe were conceived to be a set of fields – an electron field, a proton field, an electromagnetic field – and particles were reduced in status to mere epiphenomena.

In its essentials, this point of view has survived to the present day: at the fundamental level, our world is described by a set of fields, subject to the rules of special relativity and quantum mechanics plus one additional principle, the cluster decomposition [8] which requires that distant experiments give uncorrelated results. The cluster decomposition principle can be traced back to causality.

This field-theoretic approach to matter had an immediate implication: given enough energy, it ought to be possible to create material particles, just as photons are created when an atom loses energy. In 1931, Fermi used this aspect of quantum field theory to formulate a theory of the process of nuclear β decay. Throughout the 1920s it was believed that the nuclei are composed of protons and electrons, so there was no great paradox in supposing that every once in a while, one of the electrons gets out. However, in 1931, Paul Ehrenfest and Julius Robert Oppenheimer presented [9] a compelling though indirect argument that nuclei do not in fact contain electrons, and in 1932, Heisenberg proposed [10] instead that nuclei consist of protons and the newly discovered neutral particles, neutrons.[3] The mystery was, where did the electron come from when a nucleus suffered a β decay? Fermi's answer was that the electron comes from much the same place as the photon in the radiative decay of an excited atom – it is *created* in the act of decay, through an interaction of the field of the electron with the fields of the proton, the neutron, and a hypothesized particle, the neutrino.

One problem remained to be solved after 1930, in order for quantum field theory to take its modern form. In formulating the pre-field-theoretic theory of individual electrons, Dirac in 1928 had discovered (see the first work in [3]) that his equations had solutions corresponding to electron states of negative energy, that is, with energy less than the zero energy of empty space. In order to explain why ordinary electrons do not drop down into these negative-energy states, he was led in 1930 to propose [12] that almost all these states are already filled. The unfilled states, or "holes" in the sea of negative energy electrons would behave like particles of positive energy, just like ordinary electrons but with opposite electrical charge, plus instead of minus. Dirac thought at first that these "antiparticles" were the protons, but their true nature as a new kind of particle was revealed with the discovery [13] of the positron in cosmic rays in 1932.

In 1934, a pair of papers by Wendell Furry and Oppenheimer [14] and by Pauli and Victor Weisskopf [15] showed how quantum field theory naturally incorporates the idea of antimatter, without introducing unobserved particles of negative energy, and satisfactorily describes the creation and annihilation of particles and antiparticles.

◇◇◇◇◇

[3]In fact, the proton-neutron theory of nuclei was proposed by Dmitri Ivanenko [11] slightly earlier than Heisenberg. -M.S.

Pauli

Wolfgang Pauli (1910-1958) was born in Vienna to a chemist Wolfgang Joseph Pauli (né Wolf Pascheles) and his wife Bertha Camilla Schütz. His grandfather Jacob Pascheles was a well-known Hebrew publisher in Prague. He studied at the Ludwig-Maximilians University in Munich, working under Arnold Sommerfeld, Ph.D in 1921. Sommerfeld asked young Pauli to review the theory of relativity for the *Encyclopedia of Mathematical Sciences*; published as a monograph, Pauli's review remains a standard reference on the subject to this day. Pauli spent a year at the University of Göttingen as the assistant to Max Born, and the following year at the Institute for Theoretical Physics in Copenhagen, with Niels Bohr. From 1923 to 1928, he was a lecturer at the University of Hamburg. In 1928, he was appointed Professor of Theoretical Physics at ETH (Zurich) in Switzerland.

Fig. C.1 Wolfgang Pauli.

In 1938, Germany annexed Austria, and Wolfgang Pauli's Austrian citizenship was converted into German. Given his Jewish ethnic origin this created a serious problem for him in 1939. Pauli twice tried to obtain Swiss citizenship, to no avail, and eventually was forced to leave

Switzerland (in 1940) bound for the US. The only route still open to him at that time was via south of France, Spain and Portugal. Later Pauli wrote that that was an exceptionally hard journey. After his arrival in Princeton, NJ, Pauli assumed the position of Professor of Theoretical Physics at the Institute for Advanced Study.

In 1945, Pauli received the Nobel Prize in Physics for his "decisive contribution through his discovery of a new law of Nature, the exclusion principle or Pauli principle." This event resulted in an invitation from ETH offering Pauli his former position of Chair upon his return to Zurich. Surprisingly, he accepted this offer, turning down Princeton and Columbia Universities' offers. In 1946, he returned to Switzerland as an American citizen. The Swiss citizenship was granted to Pauli only in 1949.

Pauli's contributions to theoretical physics were of paramount importance. As was mentioned, he was one of the pioneers of quantum field theory. He formulated the Pauli exclusion principle, developed non-relativistic theory of spin, proposed the existence of neutrino (in analyzing β decays), invented the Pauli-Villars regularization in QFT, etc.

Wolfgang Pauli is known for his extreme perfectionism. He did not publish his results until everything was crystal clear to him. Some of his findings remained unpublished. For instance, working on the Kaluza-Klein compactification of six-dimensional gravity onto $R_4 \times S_2$, *en route* he discovered $SU(2)$ Yang-Mills theory – before the discovery of Yang and Mills. Some other Pauli's ideas and results appeared only in his letters to Niels Bohr, Werner Heisenberg and others; they were often copied and circulated by their recipients.

Jordan

Ernst Pascual Jordan (1902-1980) together with Max Born and Werner Heisenberg made fundamental contributions to quantum mechanics and field theory. But, in contradistinction with Heisenberg and Born, Jordan's name is hardly known to young people in the beginning of their physics career, let alone to the general public. You will not find his name among the winners of the prestigious prizes, although he was undoubtedly one of the strongest theoretical physicists of his generation.

Fig. C.2 Pascual Jordan. Credit: Fotothek SLUB Dresden

First, how did such a un-German name as Pascual come to Germany?

Pascual Jordan's ancestor was a Spanish nobleman and cavalry officer whose name was Jorda. After the Napoleonic wars, he settled in Hanover, which in those days was a possession of the British royal family. The family had a tradition: the firstborn son in each generation gets christened Pascual. Under the influence of the German language environment, Jorda turned into Jordan and became pronounced in German, with Jo pronounced as Yo.

Very early Pascual Jordan entered Gannover Technical University and graduated in 1921. To work on his thesis he moved to Göttingen. University of Göttingen then (in 1923) was at the height of its fame in mathematics and physics. In Göttingen Jordan almost immediately demonstrated his talent and became Richard Courant's assistant (Richard Courant was a great German mathematician who had to flee to the US in 1933 because of his Jewish origin).[4] Later he became Max Born's assistant (Max Born also had to flee from the Nazis to England in 1933).

Mid and late 1920s was the time when Jordan's scientific career reached its peak. Among other achievements discussed in the Introduction, he devised a type of non-associative algebra, now named the Jordan algebras in his honor, which are still remembered mostly by mathematicians.

At the age of 30, in 1933, Pascual Jordan entered NSDAP (the Nazi Party) and became an active supporter of Nazism in the years to come. Basically at that time, he left physics and never returned to it in earnest. Jordan found employment with the research department of Luftwaffe (German Airforce). His dream was to develop a super-weapon for Hitler. As a punishment for Jordan's former collaboration with the Jewish scientists Courant and Born, the authorities at the very top of the Nazi hierarchy never allowed Jordan to carry out important research such as the German atomic bomb project.

After Germany's defeat, Jordan had problems: at first he was not cleared by Denazification Committee and lost his job. The above Committee requested letters of recommendation from colleagues who did not taint themselves by collaborating with the Nazis. In 1948, Jordan wrote a letter to Max Born who was in Scotland at that time and asked him to write a petition to the Denazification Committee on his behalf, downplaying his (Jordan's) participation in Nazi activities, and rather emphasizing his "decision to stay in Germany to protect the Jewish contribution to physics from the racial side of Nazism." Apparently, Jordan considered the other side of the Hitler regime as appropriate.

Born politely refused by sending a letter with a list of his relatives and close friends, slaughtered by the Nazis, without any comments.

Oddly enough, Jordan's career was saved by Wolfgang Pauli who vouched for him. In 1948, Jordan was allowed to resume academic activities, and eventually, in 1953, he assumed his former position of a full professor. By that time his interests shifted, however, from science to politics. Jordan was elected to *Bundestag* (Parliament) from the Christian Democratic Union. In 1957, Jordan proposed to equip *Bundeswehr* with tactical nuclear weapons. His idea was never realized because Americans disagreed.

---

[4]Richard Courant's son, Hans Courant, was a physics professor at the University of Minnesota for many decades.

Had Jordan not joined the Nazi party in 1933, it is quite likely that he would have received the Nobel Prize in physics with Max Born in 1954. That year Born shared the Nobel Prize with Walter Bothe.

◇◇◇◇◇

Born

Of all founding fathers of quantum theory, the least known to the general public is Max Born (1882-1970). He was born in Breslau which at that time was part of Prussia (currently Wroclaw, Poland), into a wealthy family. His mother died when Max was four years old. In a few years his father remarried. In 1890, Max Born's father died too. This turmoil in his family apparently resulted in Born's almost pathological shyness and other psychological problems. In 1891, Max matriculated to the University of Breslau, where his initial exposure to higher mathematics occurred.

Fig. C.3 Max Born.

Given his interest in mathematics, which was to play a dominant role in Born's work, it was not surprising that in a few years he ended up in Göttingen University, at that time, *the* Mecca of mathematicians and theoretical physicists. At the time Born arrived there, there were three great mathematicians on the faculty: Hermann Minkowski, Felix Klein, and David Hilbert, arguably the greatest mathematician of the 20th century. Born must have shown signs of great promise because he soon became Hilbert's "scribe."

Born attended a seminar conducted by Klein on the subject of elasticity. Although not particularly interested in the subject, Born was obliged to present a paper. Klein was impressed, and invited Born to submit a thesis on the subject of "Stability of Elastica in a Plane and Space" – a subject near and dear to Klein. Born made it clear to Klein that he was not particulary fond of this topic, an admission which offended Klein. Since Klein had the power to make or break academic careers, Born felt compelled to continue his studies in elasticity. In 1906, he was awarded a prize for his results. A month later, he obtained his PhD in mathematics *magna cum laude*.

Shortly after, he was conscripted into the army for his compulsory one year service, but he was discharged much earlier because of an asthma attack, something that would re-occur throughout his life especially when he was under stress. By this time, Born came to the conclusion that he would never reach the level of Hilbert or other great mathematicians, and turned to theoretical physics. Luckily he received an invitation from Minkowski to come to Göttingen to collaborate on general relativity. This invitation was for the fall of 1908, and by next January, Minkowski died at the age of 44 because of a ruptured appendix. By this time Born had gone far enough with the research so that he could carry on alone. The work brought him an offer to become a lecturer in Göttingen.

In 1912, Born met Hedwig Ehrenberg; after a year they were married. Their marriage endured until Born's death in 1970 despite the fact that it was a troubled relationship. In 1916, Born accepted an assistant professorship in Berlin which was a promotion from his Göttingen lectureship. That's where he met Einstein, and they became friends till Einstein's death in 1955.

In 1921, Born returned to Göttingen – this time as a full professor. He managed to negotiate positions for assistants. Heisenberg was one of them. By 1923, Born's interests shifted to the quantum theory – the "old" quantum theory of Planck and Bohr. The new quantum theory began with Heisenberg the following year.

In June 1925, being in Copenhagen, Heisenberg suffered from an extreme attack of hay fever. He went to a tiny resort island called Helgoland in the North Sea to seek relief. By the time he returned to Göttingen, he had discovered the rudiments of quantum mechanics. His seminal 15-page paper entitled "On a quantum-theoretical interpretation of kinematic and mechanical relations," was published in the September 1925 issue of Zeitschrift für Physik. In early July, Heisenberg sent one copy of the paper to Pauli and gave the other to Born soliciting his advice as to whether or not to publish it.

In this paper, Heisenberg formulated all algebraic relations between observable operators which we attribute now to Heisenberg formulation of quantum mechanics. Later Heisenberg's construction grew into the construction known as "matrix mechanics," although, quite surprisingly, Heisenberg had never heard of matrices and matrix calculus in general.

In his paper, Heisenberg notes: "A significant difficulty arises, however, if we consider two quantities $X(t), Y(t)$ and ask after their product $X(t)Y(t)$. Whereas in classical theory $X(t)Y(t)$ is always equal to $Y(t)X(t)$, this is not necessarily the case in quantum theory."

Born studied Heisenberg's draft and very soon realized that he knew what this strange algebra was, that Heisenberg reinvented "matrix" multiplication, which Born had learned many years earlier in Breslau. It was Born who then derived the famous canonical commutation relation

$$[xp] = i\hbar\,,$$

a basis for all modern quantum physics, including canonic quantization in field theory.

At the time, Born experienced the mixture of physical and emotional exhaustion and decided to take a few days off to recover. He turned over what he had done to a student, Pascual Jordan. Their joint work completed the matrix mechanics; it was published in the fall of 1925 in the Zeitschrift für Physik.

In 1926, Born applied the Schrödinger wave mechanics to the problem of quantum-mechanical scattering. The approximation method he used – the "Born approximation" – is the one routinely taught in every course on quantum mechanics. Born also formulated the interpretation of the Schrödinger wave function Ψ that has been with us ever since. He postulated that the wave function was the probability amplitude for finding, say the electron, at a given position and given time. Probability itself was obtained by taking $|\Psi|^2$. This work brought him the 1954 Nobel prize.

In January 1933, the Nazis came to power in Germany. In May, Born was dismissed from Göttingen University for racial reasons. He began looking for a new job. Cambridge University offered him a three-year lectureship in mathematics which he accepted. Once this was decided, Born tried to arrange positions out of Germany for all his younger colleagues. With some he had success and with some he didn't. Most of the latter perished in Nazi Germany.

At the end of his Cambridge visiting position, Born's situation became desperate. In 1935, the Born family had their German citizenship revoked, rendering them stateless. A few weeks later Göttingen cancelled Born's doctorate. After that he accepted an offer from C. V. Raman to come to Bangalore (India) in 1935, but the Indian Institute of Science did not manage to create an additional chair for him. Then Born considered an offer from Pyotr Kapitsa in Moscow, and even started taking Russian lessons from Rudolf Peierls's Russian-born wife Genia. Fortunately, at that time Charles Galton Darwin offered him Tait Professorship at the University of Edinburgh, an offer that Born promptly accepted, assuming the chair in October 1936.

After Born's retirement in 1952, he decided to return to Germany. J. Bernstein writes [16]: "He and his wife settled in the spa town of Bad Pyrmont not far from Göttingen. Born gave as one of his principal reasons for returning to Germany, economics. He said that his pension from Edinburgh was small, and that he would be living on the pension that he now got from Göttingen. I suspect that this justification was only a part of the reason. He could have used this pension elsewhere, and he had the Nobel Prize money as well as income from his books. I think that he and his wife felt German. It was their mother tongue. Born had fought for his country in the First World War. Germany had been their home and would remain their home for the rest of their lives. Many of Born's friends and colleagues did not understand, but nostalgia for a lost homeland is a very powerful emotion. Born and his wife were buried in Göttingen. On their gravestone is carved $[xp] = i\hbar$.

◇◇◇◇◇

Dirac

Paul Adrien Maurice Dirac (1902-1984) was one of the founding fathers of quantum field theory. He discovered the relativistic equation for the electron, which now bears his name, and laid the foundation for Quantum Electrodynamics (QED). The remarkable notion of an antiparticle to each fermion particle e.g. the positron as antiparticle to the electron stems from the Dirac equation. Then he proposed and investigated the concept of a magnetic monopole. Finally, he quantized the gravitational field, and developed a general theory of quantum field theories with dynamical constraints, which forms the basis of the gauge theories and superstring theories of today. In 1933, Dirac shared the Nobel Prize in Physics with Erwin Schrödinger, "for the discovery of new productive forms of atomic theory."

Fig. C.4 Paul Dirac.

Paul Dirac was born in Bristol, England into a family of a teacher and a librarian. His father was of Swiss descent, his mother tongue was French. Dirac studied electrical engineering at the University of Bristol.

Shortly before he completed his degree in 1921, he passed the entrance examination for St John's College, Cambridge, and was awarded a GBP70 scholarship (approximately $ 5,000 in 2015 dollars). This was not enough to live and study at Cambridge. Despite his having graduated with a Bachelor of Science degree in engineering, the economic climate of the post-war depression was such that he was unable to find work as an engineer. Instead, he took up an offer to study for a Bachelor of Arts degree in mathematics at the University of Bristol free of charge.

In 1923, Dirac graduated, once again with first class honors, and moved to Cambridge. There, Dirac pursued his interests in the theory of general relativity and in the nascent field of quantum physics. He completed his PhD in June 1926 with the first thesis on quantum mechanics to be submitted ever. He then continued his research in Copenhagen and Göttingen.

Working on quantum theory Dirac noticed an analogy between the Poisson brackets of classical mechanics and the recently proposed quantization rules in Heisenberg's matrix formulation of quantum mechanics. This observation allowed Dirac to obtain the quantization rules in a novel and more illuminating manner. Twenty years later, Richard Feynman was inspired by this work which eventually led him to a revolutionary path-integral formulation of field theory.

The world of physicists is small. In 1934, when Paul Dirac was visiting Princeton, NJ, he met Margit Wigner, Eugene Wigner's sister. Margit, known as Manci, came from her native Hungary to see her brother. While at dinner at the Annex Restaurant she noted a "lonely-looking man at the next table." In 1937, Margit and Paul married. Dirac adopted Margit's two children, Judith and Gabriel. Paul and Margit Dirac had two children together, both daughters, Mary Elizabeth and Florence Monica.

After his retirement from Cambridge in 1962, and two years lecturing in New York, Dirac relocated to Florida to be near his elder daughter, Mary. There, Dirac spent his last years of both life and physics research at the University of Miami in Coral Gables, Florida, and Florida State University in Tallahassee, Florida.

Dirac was known among his colleagues for his remarkable personality. Albert Einstein said of him, "This balancing on the dizzying path between genius and madness is awful." His friends and colleagues were well aware of Dirac's precise and taciturn nature. Cambridge physicists jokingly defined a unit of a "dirac," which was one word per hour. When Niels Bohr complained that he did not know how to finish a sentence in a scientific article he was writing, Dirac replied, "I was taught at school never to start a sentence without knowing the end of it." He criticized J. Robert Oppenheimer's interest in poetry: "The aim of science is to make difficult things understandable in a simpler way; the aim of poetry is to state simple things in an incomprehensible way. The two are incompatible."

An anecdote recounted in a review of the 2009 Dirac's biography tells of Werner Heisenberg and Dirac sailing on an ocean liner to a conference in Japan in August 1929. "Both still in their twenties, and unmarried, they made an odd couple. Heisenberg was a ladies' man who constantly flirted and danced, while Dirac – an Edwardian geek, as biographer Graham Farmelo puts it – suffered agonies if forced into any kind of socializing or small talk. 'Why do you dance?' Dirac asked his companion. 'When there are nice girls, it is a pleasure,' Heisenberg replied. Dirac pondered this notion, then blurted out: 'But, Heisenberg, how do you know beforehand that the girls are nice?'"

Another story told of Dirac is that when he first met the young Richard Feynman at a conference, he said after a long silence, "I have an equation. Do you have one too?"

After he presented a lecture at a conference, one colleague raised his hand and said "I don't understand the equation on the top-right-hand corner of the blackboard." After a long silence, the moderator asked Dirac if he wanted to answer the question, to which Dirac replied "That was not a question, it was a comment."

Dirac was also noted for his personal modesty. He called the equation for the time evolution of a quantum-mechanical operator, which he was the first to write down, the "Heisenberg equation of motion." Most physicists speak of Fermi-Dirac statistics for half-integer-spin particles and Bose-Einstein statistics for integer-spin particles. While lecturing later in life, Dirac always insisted on calling the former "Fermi statistics". He referred to the latter as "Einstein statistics" for reasons, he explained, of "symmetry."

◇◇◇◇◇

Heisenberg

Werner Karl Heisenberg (1901-1976) was a German physicist who made a major contribution to quantum theory. In 1932, Heisenberg was awarded the Nobel Prize in Physics "for the creation of quantum mechanics."

Prof. Dr. Werner Heisenberg,
der erst 32jährige Physiker an der Universität Leipzig, erhielt für seine Arbeiten auf dem Gebiet der Quantentheorie die Hälfte des Nobelpreises für Physik aus dem Jahre 1932
Phot. Max Löhrich

Fig. C.5 Werner Heisenberg in 1933.

Heisenberg was born in Würzburg, Germany, to Kaspar Earnesta August Heisenberg, a secondary school teacher of classical languages and his wife, Annie Wecklein. He studied physics and mathematics from 1920 to 1923 at the University of Munich and later at Göttingen. At Munich, he studied under Arnold Sommerfeld while at Göttingen, he studied physics with Max Born and James Franck, and mathematics with David Hilbert. Créme de la créme! He received his doctorate in 1923.

From 1924 to 1927, Heisenberg was a *Privatdozent* (Associate Professor) at Göttingen. It was in Copenhagen, in 1927, that Heisenberg developed his uncertainty principle, while working on the mathematical foundations of quantum mechanics.

In 1927, Heisenberg was appointed *ordentlicher Professor* of theoretical physics and head of the department of physics at the Universität Leipzig. In 1928, Heisenberg laid the foundation of the quantum theory of (anti)ferromagnets, the so-called Heisenberg spin chain models. In early 1929, Heisenberg and Pauli submitted the first of two papers laying the foundation for relativistic quantum field theory. Shortly after the discovery of the neutron by James Chadwick in 1932, Heisenberg submitted a paper on the neutron-proton model of the nucleus. He was the first to introduce isospin to explain symmetries of the then newly discovered neutron.

After Adolf Hitler came to power in 1933, Heisenberg was attacked in the press as a "White Jew" by the Deutsche Physik (German Physics) movement for his insistence on teaching theories developed by Jewish scientists, such as special and general relativity. As a result, he came under investigation by the SS.[5] The issue was resolved in 1938 by Heinrich Himmler, head of the SS: Heisenberg was rehabilitated to the physics community during the Third Reich.

In mid-1936, Heisenberg presented his theory of cosmic ray showers. In January 1937, Heisenberg met Elisabeth Schumacher at a private music recital. Elisabeth was the daughter of a well-known Berlin economics professor. Heisenberg married her on April 29, 1937. Fraternal twins Maria and Wolfgang were born in January 1938, whereupon Wolfgang Pauli congratulated Heisenberg on his "pair creation", a word play on a process from elementary particle physics, pair production. They had five more children over the next 12 years: Barbara, Christine, Jochen, Martin and Verena.

Heisenberg enjoyed classical music and was an accomplished pianist.

In 1939, shortly after the discovery of nuclear fission, the German nuclear project, also known as the *Uranverein* (Uranium Club), was launched. Heisenberg was the head of research and development in the project.

[5]This investigation found no traces of Jewish blood in Heisenberg's family history.

After WWII, in the last two decades of his research career, Heisenberg spent most of his time exploring a possible version of a unified field theory. These efforts did not give rise to further developments.

Considerable controversy surrounds Heisenberg's work on the Nazi atomic bomb project during World War II.

From the early postwar years till 2002 most historians of science followed a theory according to which the team led by Heisenberg deliberately sabotaged the German nuclear project, accounting for its failure. This myth can be traced back to a letter Heisenberg sent to Robert Jungk, see [17]. Thomas Powers, who wrote [21] the monumental treatise *Heisenberg's War*, was an ardent proponent of this theory.

What changed in 2002?

It was known that in 1941, Heisenberg visited Bohr in Copenhagen. They had a private conversation. No details of this conversation were available to historians of science, which gave rise to unlimited speculations. In 2002, a letter written by Bohr to Heisenberg around 1957, sealed in the Bohr personal archive, was made available to the public. To my mind, this letter leaves no basis for the myth of sabotage. This conclusion seems to be shared by historians in more recent publications. In particular, Robert Jungk himself no longer stands behind the interpretation of his 1956 book. Moreover, he accused [22] von Weizsäcker, whose interviews were a source for his book, of misleading him, and Heisenberg of confirming von Weizsäcker's claims:

> I accuse a few [eyewitnesses], who even today are respected by the public, of conscious distortion of history by means of deliberate misrepresentations and – I do not even hesitate to use this devastating word – the occasional lie. [...] It was Carl Friedrich von Weizsäcker who described to me very forcefully that the German scientists did not want to build the atomic bomb. At that time he used the expression "passivists" to describe this circle of people.

Because of its importance for the history of science, I quote here Bohr's letter [23] in full:

> Dear Heisenberg,
>
> I have seen a book, "Stærkere end tusind sole" ["Brighter than a thousand suns"] by Robert Jungk [17], recently

published in Danish, and I think that I owe it to you to tell you that I am greatly amazed to see how much your memory has deceived you in your letter to the author of the book, excerpts of which are printed in the Danish edition.

Personally, I remember every word of our conversations, which took place on a background of extreme sorrow and tension for us here in Denmark. In particular, it made a strong impression both on Margrethe and me, and on everyone at the Institute that the two of you spoke to, that you and Weizsäcker expressed your definite conviction that Germany would win and that it was therefore quite foolish for us to maintain the hope of a different outcome of the war and to be reticent as regards all German offers of cooperation. I also remember quite clearly our conversation in my room at the Institute, where in vague terms you spoke in a manner that could only give me the firm impression that, under your leadership, everything was being done in Germany to develop atomic weapons and that you said that there was no need to talk about details since you were completely familiar with them and had spent the past two years working more or less exclusively on such preparations. I listened to this without speaking since [a] great matter for mankind was at issue in which, despite our personal friendship, we had to be regarded as representatives of two sides engaged in mortal combat. That my silence and gravity, as you write in the letter, could be taken as an expression of shock at your reports that it was possible to make an atomic bomb is a quite peculiar misunderstanding, which must be due to the great tension in your own mind. From the day three years earlier when I realized that slow neutrons could only cause fission in Uranium 235 and not 238, it was of course obvious to me that a bomb with certain effect could be produced by separating the uraniums. In June 1939 I had even given a public lecture in Birmingham about uranium fission, where I talked about the effects of such a bomb but of course added that the technical preparations would be so large that one did not

> know how soon they could be overcome. If anything in my behavior could be interpreted as shock, it did not derive from such reports but rather from the news, as I had to understand it, that Germany was participating vigorously in a race to be the first with atomic weapons.
>
> Besides, at the time I knew nothing about how far one had already come in England and America, which I learned only the following year when I was able to go to England after being informed that the German occupation force in Denmark had made preparations for my arrest.
>
> All this is of course just a rendition of what I remember clearly from our conversations, which subsequently were naturally the subject of thorough discussions at the Institute and with other trusted friends in Denmark. It is quite another matter that, at that time and ever since, I have always had the definite impression that you and Weizsäcker had arranged the symposium at the German Institute, in which I did not take part myself as a matter of principle, and the visit to us in order to assure yourselves that we suffered no harm and to try in every way to help us in our dangerous situation.
>
> This letter is essentially just between the two of us, but because of the stir the book has already caused in Danish newspapers, I have thought it appropriate to relate the contents of the letter in confidence to the head of the Danish Foreign Office and to Ambassador Duckwitz.

From this letter it seems clear that Heisenberg's team worked in earnest to make the bomb. They failed not because they sabotaged the project, but because they were not qualified to solve the problems that arose in the course of their work. I mentioned already a fatal mistake in Bothe's experiment (see footnote 120 on page 233). As for the critical mass calculation, this issue is discussed in great detail in Jeremy Bernstein's paper "Heisenberg and the Critical Mass" [24]. I will try to summarize the main points of this paper as follows.

The 1939 wartime report [written by Heisenberg] is called "Die Möglichkeit der technischen Energiegewinnung aus der Uranspaltung" – "The possibility of the technical use of the energy gained from uranium fission," see [25]. In this report Heisenberg derives a rather crude formula for the critical mass. No numerical estimates were presented. Using the Heisenberg formula, one can obtain the critical radius and the corresponding mass, 31 cm and approximately one ton, respectively.[6] These numbers agree with the first estimate Heisenberg presented to his colleagues at Farm Hall (see below). In 1940, Karl Wirtz heard Heisenberg commenting on this calculation.

The Farm Hall was bugged so that conversations of all detainees were recorded and transcribed. On August 6, 1945, the detained German physicists learned that a new weapon had been dropped on Hiroshima. They did not believe that it was nuclear.

Fig. C.6 Farm Hall, circa 1945. Ten German physicists – Erich Bagge, Kurt Diebner, Walther Gerlach, Otto Hahn, Paul Harteck, Werner Heisenberg, Horst Korsching, Max von Laue, Carl Friedrich von Weizsäcker, and Karl Wirtz – were interned here from July 3, 1945, to January 3, 1946.

[6] That is, approximately 50 times larger than the actual value of the critical mass.

When they finally were persuaded that it was, they began trying to explain it. That evening, Otto Hahn and Heisenberg had a conversation. Heisenberg gave Hahn an estimate based on the data concerning the Hiroshima explosion published in newspapers. Heisenberg reasoned as follows. He knew that the Hiroshima explosion was about equivalent to 15,000 tons of TNT, and he knew that this amount corresponded to the fission of about 1 kg of uranium. Then he estimated that this would require about eighty generations of fissions assuming that two neutrons are emitted per fission. He then assumed that during this process the neutrons flow out to the boundary in a random walk of eighty steps with a step length equal to the mean free path for fission. This gave him a critical radius of 54 cm and a critical mass of several tons. (The correct estimate would give 15-20 kg.)

On August 14, 1945, Heisenberg gave a lecture, on this subject at Farm Hall to the other nine detainees. From his August 14 lecture, one can infer that he had finally understood the basics of the problem. This time Heisenberg derived a reasonable critical mass of 15 kg.

Then Bernstein continues:

> There is one especially surreal aspect of this discussion that took place after the second bomb was dropped on Nagasaki. The mass of material for this bomb was given in news reports and it seemed too small. The Germans indulged in all sorts of wild speculations as to why this was so. It never occurred to them that the Nagasaki bomb was made of plutonium, despite the fact that von Weizsäcker, who had introduced the idea of transuranics into the German program, was in the audience.
>
> To have admitted that plutonium was used was to admit that the Allies had a vast reactor development and that everything the German scientists had worked on for so long and so hard had been insignificant. Heisenberg's lecture, which represented the high water mark of the German understanding of nuclear weapons, shows that in the end they understood very little.
>
> Prior to Hiroshima the Germans were absolutely convinced on the basis of their own experience that a nuclear bomb could not be built in the immediate future. Their belief was based on the idea of their superiority: they

were absolutely convinced that they were ahead of everyone else in their study of nuclear chain reaction. Because they had not been able to build a nuclear reactor, they were sure that no one else had done so. These points are made very explicitly in the Farm Hall transcripts.

References

[1] S. Weinberg, "The Search for Unity: Notes for a History of Quantum Field Theory," *Discoveries and Interpretations: Studies in Contemporary Scholarship, Volume II* , in Daedalus, Vol. 106, No. 4, (Fall, 1977), pp. 17-35. S. Weinberg, "What is quantum field theory, and what did we think it is?" Proceedings of Symposium and Workshop Conceptual foundations of quantum field theory, Boston, USA, March 1-3, 1996 in *Conceptual Foundations of Quantum Field Theory*, Ed. Tian Yu Cao, (Cambridge University Press, 1999) [hep-th/9702027].

[2] M. Born, W. Heisenberg and P. Jordan, Z. Phys. **36**, 336 (1926), Sec. 3.

[3] P. A. M. Dirac, Proc. Roy. Soc. (London) **A117**, 610 (1928); *ibid.*, **A118**, 351 (1928); *ibid.*, **A126**, 360 (1930).

[4] M. Born and P. Jordan, Z. Phys. **36**, 336 (1926).

[5] P. A. M. Dirac, Proc. Roy. Soc. (London) **A112**, 661 (1926), Sec. 5.

[6] P. Jordan and E. Wigner, Z. Phys. **47**, 631 (1928).

[7] W. Heisenberg and W. Pauli, Z. Phys. **56**, 1 (1929); *ibid.* **59**, 168 (1930).

[8] E. H. Wichmann and J. H. Crichton, Phys. Rev. **132**, 2788 (1963).

[9] P. Ehrenfest and J. Oppenheimer, Phys. Rev. **37**, 333 (1931).

[10] W. Heisenberg, Z. Phys. **77**, 1 (1932).

[11] D. D. Iwanenko, *The neutron hypothesis*, Nature **129**, 798 (1932).

[12] P. A. M. Dirac, Proc. Roy. Soc. (London) **A126**, 360 (1930); Proc. Camb. Phil. Soc. **26**, 361 (1930).

[13] C. D. Anderson, Science **76**, 238 (1932); Phys. Rev. **43**, 491 (1933).

[14] H. Furry and J. R. Oppenheimer, Phys. Rev. **45**, 245 (1934).

[15] W. Pauli and V. Weisskopf, Helv. Phys. Acta **7**, 709 (1934).

[16] Jeremy Bernstein, Am. J. Phys. **73**, 999 (2005).

[17] Robert Jungk, *Heller als tausend Sonnen. Das Schicksal der Atomforscher*, (Stuttgart, 1956), in German.

[18] *Oral History Transcript Program*, American Institute of Physics (AIP), http://www.aip.org/history-programs

[19] *The Recollections of Eugene P. Wigner as told to Andrew Szanton*, (Plenum Press: N.Y. and London, 1992), p. 241.

[20] F. Hernek, "Eine alarmierende Botschaft," *Spektrum*, Januar 1976, pp. 32-34, in German.

[21] Thomas Powers, *Heisenberg's War: The Secret History of the German Bomb*, (Knopf, 1993).

[22] Mark Walker, "Physics and propaganda: Werner Heisenberg's foreign lectures under national socialism," *Historical Studies in the Physical and Biological Sciences*, Vol. 22, No. 2 (1992), pp. 339-389.

[23] Niels Bohr Archive, Draft of letter from Bohr to Heisenberg, never sent. In the handwriting of Niels Bohr's assistant, Aage Petersen.

[24] J. Bernstein, *Heisenberg and the critical mass*, American Journal of Physics **70**, 911 (2002).

[25] W. Heisenberg *Gesammelte Werke/Collected Works*, Eds. W. Blum, H. P. Dürr, and H. Rechenberg (Springer, Berlin, 1989), Ser. A, Pt. II. p. 378.

Index

Printed in the USA
CPSIA information can be obtained
at www.ICGtesting.com
LVHW081652091123
763115LV00084B/632